susan bordo

the male body

SUSAN BORDO holds the Otis A. Singletary Chair in the Humanities at the University of Kentucky, where she is also a professor of English and women's studies. She has written and edited several books, including *Unbearable Weight: Feminism, Western Culture, and the Body*, nominated for the Pulitzer Prize, and *Twilight Zones: The Hidden Life of Cultural Images from Plato to O.J.*

the male body

the
male

A NEW LOOK AT MEN

susan

FARRAR, STRAUS AND GIROUX

body

IN PUBLIC AND IN PRIVATE

bordo

NEW YORK

Farrar, Straus and Giroux
19 Union Square West, New York 10003

Copyright © 1999 by Susan Bordo
All rights reserved
Distributed in Canada by Douglas and McIntyre Ltd.
Printed in the United States of America
Designed by Jonathan D. Lippincott
First published in 1999 by Farrar, Straus and Giroux
First paperback edition, 2000
2 3 4 5 6 7 8 9 10
Library of Congress Cataloging-in-Publication Data
Bordo, Susan, 1947–
    The male body : a new look at men in public and in private / Susan
Bordo.
        p.   cm.
    Includes bibliographical references.
    ISBN 0-374-52732-6 (pbk.)
    1. Men.   2. Masculinity.   3. Men in popular culture.
    4. Masculinity in popular culture.   5. Body, Human—Social aspects.
    I. Title.
HQ1090.B67   1999
    305.31—dc21   99-25386

Every effort has been made to secure permission to reproduce material under copyright. Grateful acknowledgment is made for use of the following:

Esquire magazine cover used by permission of *Esquire* magazine. © Hearst Communications, Inc. Also, *Esquire* is a trademark of Hearst Magazines Property, Inc. All rights reserved.

In Style magazine spread used by permission of *In Style* magazine. *In Style* is a registered trademark of Time Inc., used with permission.

Men's Fitness cover used by permission of *Men's Fitness* magazine.

"Style Points" spread used by permission of *Black Men* magazine.

Tom of Finland image provided by the Tom of Finland Foundation under international copyright law, P.O. Box 26658, Los Angeles, CA 90026, (213) 250-1685.

For Yosh and Eddie

contents

the male body

prologue

MY FATHER'S BODY

My father was prone to dandruff, and to vagrant patches of dry, flaky skin—at the thick edge of his eyebrows, around the side of his nose, behind and inside his ears. I don't know when I first noticed that my skin was beginning to pattern in exactly the same way, but it certainly wasn't until I was into my forties. When I did notice, it came as a shock of startlingly precise, irrefutable linkage between him and me. I was ambivalent about the discovery. It was not a glamorous trait to have in common with him. But it was proof of our genetic connection—a sense of belonging hard to come by in our scattered, nonreproducing family—and it made me feel close to him in some new way.

My father had always been extremely private about his body. I never saw him naked, not once. To this day, I have no idea what his unclothed buttocks looked like, or his penis. I rarely saw him with his shirt off. Nor was he emotionally revealing, except when he was angry. His hugs, although expansive and affectionate, did not linger, seemed perfunctory. While he was still working, he was constantly on the move; after he retired, he was usually sunk in a private space, too depressed to connect. That shared malady, our dry skin, felt like intimacy.

My father's body had three incarnations for me. The first belongs entirely to old photographs and stories of adventures that preceded my birth, photos which show my father well built and dashing, as hand-

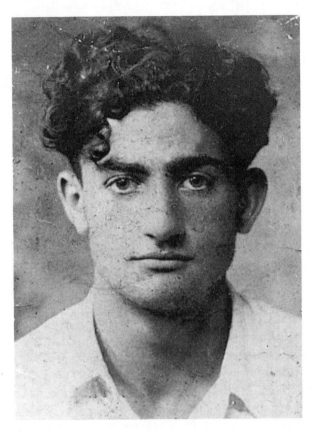

some as John Garfield, the tough Jewish street kid who became a Hollywood star. A trim and dashing young man stands beside a propeller
plane; the strong eros he projects—his windswept pose, his jaunty
sweater—is startling to me, even now that I know (as I did not as a
child) that people can have several lives. In another photo, a sensual
young man stares at me, his thick black hair and sultry, ethnic looks of
the kind that are undeniably in vogue today. The girls of his Brooklyn
neighborhood, lacking those cultural models, called him "the Jewish
John Barrymore."

   During his Brooklyn youth, my father had been popular and successful. He was a track star and a straight-A student, but somehow
managed—through his keen iconoclastic wit, feisty spirit, and willingness to take a dare—to win the respect and affection of members of the

Jewish Mafia and their "dolls," with whom he hung out on occasion. I grew up hearing stories of his romantic exploits and brushes with danger.

Many years later, when my father retired, he committed some of those stories to poetry. "He lived in Brooklyn," he wrote, "but in spirit he dwelt in King Arthur's Court." "He dreamed when knighthood was in flower, And he would rescue every hour on the hour, The fair maiden from the dark and evil tower." My father's romanticism, however, was of the Damon Runyon variety. One of the maidens with whom he fell in love and whom he tried to rescue was gangster moll Alice:

> A bobbed hair brunette, with bright violet eyes,
> Alice looked out at the world with seeming surprise.
> She was no ways naive, when you realize,
> She was a child of the gutter, sharp, and streetwise.
>
> They sent her twice to the juvenile hall,
> She escaped a second time, over the wall.
> At eighteen, she was contracted by the Brooklyn mob,
> Along with an accomplice, some underworld slob.
>
> She would sit in a parked car, on the passenger's side,
> While in the back, her partner would hide.
> She would wink and flirt, with a male passerby,
> Who would stop to inquire, with high hopes in his fly.
>
> He succumbed to temptation and poked in his head,
> And was promptly kayoed, by a pipe made of lead.
> They rolled him expertly, and where he sought romance,
> He was lucky indeed, if they left him his pants. . . .

One day, Alice unfortunately tried her charms on a plainclothes policeman, and "got three to five." On release from jail, she met my father. The two "made love everywhere, despite the summer heat, always avoiding the cop on the beat." But the mob wanted Alice back in their ranks. Informed that they're coming for Alice at eight, my father orders a gun from "Ike, the Toad" and prepares for confrontation (just a bluff, he assured me, as I listened to the tale in childhood, rapt). When the "limo pulled up at the appointed time, inside it two hoodlums well-known to crime," the Toad still has not arrived. My father is beside himself, but determined to do the manly thing. "She stays where she is!" he declares. Suddenly, the Toad arrives, out of breath, and thrusts a crumpled kerchief into my father's hands. Inside it is the gun, in pieces, unassembled. The gangsters in the car, "laughing as if their sides would split, you'd think they were watching a vaudeville skit" (which my father was by now doing too, as he got to this part of the story), decide to leave my father and Alice alone. My father and Alice finally split up after one night when my father, who had taken Alice up on the tenement roof ("to rendezvous with blankets and whiskey, 100 proof"), falls asleep, then wakes up to see her walking along the edge of the roof stark naked. That's too much excitement, even for him.

Did these incidents actually happen? Told to me by a plump, balding forty-five-year-old, they belonged to a time and place as distant as Camelot, and as unavailable for factual confirmation. (My mother never disputed my father's accounts, but she never confirmed them either.) When men did "wrong" by his wife and daughters, however, my father's boyhood skills could resurface, offering their own kind of proof. An abusive gym teacher was told in no uncertain terms that if he called me "fatty" one more time he was going to have the shit kicked out of him. The summer after my first year at college—a year in which my main accomplishment was that I had "gone all the way" on my eighteenth birthday—I dared to close my bedroom door with a boy inside the room. He was the boy with whom I had celebrated my birthday, and my father had suspicions (which I subtly encouraged) that I had come home changed; he pounded on the door, threatening to break it down if we did not come out. These scenes left me confused about what I admired and wanted in a man. It was embarrassing for a modern girl to have her father behave like a caveman. But these were the proofs of love my father offered, and they had an archetypal resonance that I couldn't deny.

My father, who had thrived in his dual role of promising scholar and street-smart adventurer, was let down by the more sober, unromantic requirements of "good provider" manliness. The Depression forced him to abandon his dreams of college and a career in journalism; after the war, he returned from the South Pacific to a wife and small daughter (my older sister) and a job as a poorly paid employee of wealthier relatives, selling candy on the road for the family business. The job was not entirely unrewarding. He enjoyed making sales and loved traveling to exotic places like New Orleans, where he ate and drank in fancy restaurants, swapping stories with the pleasure-loving southern brokers who reveled in "Yosh's" warmth and wit. He'd send me post-cards from the "Fabulous White Way" of the Las Vegas Strip; "Greet-ings from the land of lost wages, daughter!" he wrote on one in his distinctive, jaunty handwriting. But coming home from these trips was always to return to his subordinate status in the company, and what he increasingly came to experience as a failed life. He identified with Willy Loman.

This is the father that I knew through most of my growing-up life: the balding, large-bellied salesman with his sample cases and love of Chinese food, his generosity and intelligence like sunlight and his sudden dark furies. We three girls (and my mother) basked in the sunlight—the jokes, the stories, the trips away from Newark, into New York City—and cringed when the clouds came out. My mother seemed always to be waiting, smoking or dozing in my father's armchair, putting canned goods away, feeding the cat, until called upon to spring into action on his return. There would be tiny hotel soaps, plastic dolls dressed in buckskin, pecan pralines, restaurant matchbooks to add to my collection. The thrill of suitcases opening. Whisking away for Chinese food, to an air-conditioned restaurant. We were all in a good mood for perhaps a few hours; then, inevitably, came the crash. Usually, it began with a petty squabble, among the kids. Always, it escalated to something global, between our parents, and then metaphysical, between my father and God. "Why can't you ever . . . ?" "Why do I have to put up with . . . ?" "What did I do to deserve . . . ?"

A photograph from this period in my father's life shows him at a table with three other men in business suits, presumably at a candy convention. There are ashtrays and a Heineken bottle on the white tablecloth. Someone must have been making a presentation; the three other men are smiling, one laughing. My father has his eyes closed, cigar in mouth, hands resting on the table, five fingertips of one hand touching the five fingertips of the other, almost as if he were in prayer. It looks like a composite photograph (although it isn't); my father's presence among the other salesmen is like a finger dislocated from the rest of a hand. I'd often seen him the same way at family gatherings, slightly apart, absorbed in some inner musings. My father of lost dreams, of the secret unlived life.

As the years passed, he became more brooding. He sought proofs of respect. Once, at age twenty-nine, I made the mistake of requesting that my father smoke his cigar outside the airless basement apartment I was living in as a graduate student. He blew up at me with a fury that surpassed the (considerable) furies I had witnessed throughout my childhood. Love me, love my cigar. My father sometimes said this jokingly, but we all knew he meant business. Rarely physically abusive, he could nonetheless be verbally vicious and frighteningly expressive. His

eyes would darken and his mouth would set, almost in a snarl; often, he would leave rooms and houses, slamming the door behind him and managing to convey, with impressive believability, the possibility that he might never return. (He always did, later the same day or evening.)

As my father became progressively more frail and dependent on others, he became a "different man." At least, that was the way his second family—the children and grandchildren of my stepmother—experienced him. After he died, they were shocked and unbelieving to hear our stories of his petty tyrannies and temper tantrums. The man *they* knew was gentle "Grandpa Yosh," who loved children and didn't have a mean bone in his body. I think they believed that we, his feminist daughters, were remembering the past falsely through the screen of our own unresolved resentments and angers. But my father always *had* been sweet, there was never any question of that. He had been sweet, and he had also been armored, ready to be betrayed and treated without respect, ready for a fight as he had learned to be from his brawling father, from the screen heroes whom he adored, and from the minor gangsters he caroused with on the streets of Brooklyn. Ready to lay down the law and walk out the door when he was hurt rather than tell you how sad he felt, how sorry.

My mother had known this, long before I did. She forbade us to crit-
icize my father's habits in any way. I saw this, not incorrectly, as an aspect of his dominance of our lives. When he sat down at the television
to watch a baseball game, volume blaring, cigar smoke wafting across
the room (or a wet, extinguished butt in the ashtray, sending off its
acrid fumes), it seemed that he owned our collective space absolutely—
unconsciously, yet absolutely. He owned the sights, and sounds, and
(this was the worst, as far as my sisters and I were concerned, covertly
grimacing at each other, our noses pinched between our fingers) the
smells. Even today, I can't listen to the sounds of a televised sports
event without feeling irritated, and vaguely queasy.

I wasn't wrong about my father's privilege in our household. But
apart from the stories he told me, I knew very little, in those days,
about the rest of my father's life. My mother did. She had known the
streetwise, book-smart dazzler with bedroom eyes. She had known the
young husband who sent long, romantic letters home from his ship,
quoting Keats, Shelley, and Byron. She knew that my father's cigars
were like pacifiers to him, portable reassurances that despite its many
griefs and disappointments life could still be sweet and satisfying. She
knew that when he yelled at us for complaining about the smoke, he
was actually horrified at the thought that something he needed so
badly might be disgusting to those around him. She knew that his
seeming callous obliviousness in the face of our gagging throats and
burning eyes was a defiant defense against his own deep shame.

My father's body as I knew it in my childhood was, on the surface, the
very opposite of what was to reveal itself as my romantic ideal. Continually drawn later in life to tall, slender, aristocratic Protestants, I
seemed to be fleeing, with a suspect desperation, everything my father
was. But the surface of the body can mislead. The poise and politeness
of my WASPs often kept them just as impenetrable and remote from
me as my father's armor of Jewish affability and public largesse. And
then, too, dotting my sexual landscape, there were always the *other*
boys, the reckless, volatile ones with dark hair and sturdy legs. The
first time I slept with a man whose body was solid, it was like discov-
ering some deeply repressed but incontrovertible bodily need. I found
his hands—the hands of a worker, a peasant (although he himself was

a student of philosophy)—extremely moving. I savored looking at them the way I imagine men look at the cultural emblems of femininity, magnetized by meanings over which they have little control. My father's last body is the one I was most intimate with. In his last weeks, his face became very thin, with sunken cheeks that brought out the Russian-Jewish structure of his face. He looked very little like the father I had known most of my life, more like an ancestral photo of a frail, aged relative from "the old country." I felt that his body was joining his mother's and father's, as my own body had been changing, joining his through resemblances that had been muted by a temporary and fragile individuation.

My relatives had warned me, as I got ready to drive down for what I knew would be a final visit, that his appearance might be shocking and upsetting to me. What I was not prepared for was the deep comfort—perhaps it could even be called pleasure—that I got from simply being alone with him, close to his body, from holding his hand or touching his shoulder as long as I wanted to, from looking at him with such an unobstructed intimacy of gaze, from lingering with him and over him. Exhausted, cocooned in a state of consciousness that was neither sleep nor waking, he mostly merely lay there, accepting my physical closeness, my caresses. Just once, he spoke to me. He told me he was cold, and he asked me to cover him. When I did, he thanked me with the grace of a courtier, in the affectionate male lingo of his Runyonesque Brooklyn: "Thank you, doll. Thank you, doll." As wretched as he looked, I feasted on the sight of him as if he were my infant boy, and it was very hard to leave his side.

private parts

in hiding and on display

in the dark with men's bodies

Becky Stone was the first of my friends to actually *see* one. We were fifteen, and Becky was going steady with a nice boy who "stopped" when asked, so she was allowed certain explorations too dangerous for the rest of us. I was incredibly jealous of the liberties she enjoyed. But I was also terrified of "having" to look at it myself.

My direct acquaintance, up to that point, had been entirely tactile, furtive, and fragmentary, like the blind man's picture of the elephant. I only knew what came in contact with me in the dark; the rest was left to my imagination, which often got things wrong. From Jeffrey Schwartz, pressing himself against me as we danced in my basement, I had learned that the thing could become hard, and was capable of asserting itself at a surprising angle from the body. I was twelve; from that moment on I imagined that my science teacher, whose pants had an odd crease in front—probably caused by a wallet—was perpetually erect.

Then I was fourteen, and Bobby Cohen was lying heavily on top of me in someone else's basement, his nose running. He did not seem to be having a good time as he humped away. He had a bad cold, and I'm sure would rather have been home (as I wished I was); but something was at stake here more important than pleasure, clearly, for both of us. We were both fully clothed; I stuck my hand in between his body and mine (but over his clothes) and pressed it against him, if only to take an intermission from the incessant humping. A sharp thrill went through me, which I decided to ignore, put on hold for another time. I knew that I could only absorb so much knowledge from this new realm, or I might be scared off forever.

Becky was more relaxed about these things than I. When she finally gave in—her boyfriend had been pleading with her for weeks to "take it out" of his pants—she came back shrieking, giggling, with details that inspired horror, awe, revulsion. "It shoots straight up from a bunch of hair!" That seemed wrong, unnatural, like a particularly ugly mixed metaphor or a mythological beast whose parts were mismatched. I shuddered at the thought that someday I would have to look at one too. But actually, the opportunities for that were minimal growing up in the late fifties and early sixties, at least at my high school, where most heterosexual exploration was organized around forays, skirmishes, invasions into the territory of *our* bodies—women's bodies.

"Did you take it off?" "Did you see it?" "Did you touch it?" Boys compared notes after dates. To become a man, a boy went through a certain education (however flawed or distorted) in women's bodies. Our panties were skins to be peeled, the fasteners on our bras latches to be unlocked; getting underneath or inside might have been for purposes of conquest or "scoring" but it also gave information. My girlfriends and I were so preoccupied with the proper management of this male education (for it seemed to be in our hands) that we often forgot about our own.

Was I odd to have reached puberty without having seen a penis? It is true that my father was especially modest, and I suspect that my friends with brothers at home were more knowledgeable than I. But the vaguely repellent mystery of my father's body—what horrible thing was lurking down there, under his baggy boxer shorts?—was not solved for me by looking around the world outside my family. The fact is that although my father's prissiness about his body may have been extreme among actual men, it seemed to be perfectly mirrored in the attitudes, symbols, stories, and images of the culture around me.

It caused a furor in 1961 at Mattel, Inc., when female executives argued that Barbie's new partner Ken ought to have a bulge in his groin. Barbie's own breasts, if translated into human proportions, would have made her Jayne Mansfield. She had been modeled almost exactly on the German "*Bild* Lilli" doll, originally marketed as a mini sex toy for adult men: big breasts, cinched waist, long legs, and slightly lascivious face. Her biographer, M. G. Lord, describes her body as "*la différence* incarnate." But try to endow Ken with *his* "difference incarnate" and all hell broke loose. The designers had to try out three versions of Ken's crotch in an effort to appease nervous male executives. Charlotte Johnson, Barbie's clothing designer, recalls: "One was—you couldn't even see it. The next one was a little bit rounded, and the next one really *was*. So the men—especially one of the vice presidents—were terribly embarrassed. So Mrs. Handler and I picked the middle one as being the one that was nice-looking. And he said he would never have it in the toy line unless we painted Jockey shorts over it."

Some might argue that the squeamish executives only wanted to protect innocent children from a too sexually explicit plaything. After all, these dolls weren't designed as a course in sex instruction. But we're not talking here about testicles, shaft, head—just a little plastic mound. And if keeping sexual messages muted was the issue, then how did Barbie get away with being a bosomy vamp while even an anatomically vague allusion to Ken's sexuality was so problematic?

If one didn't get to see one at home or in the backseat of a car, where *did* a girl get to see one? Museum statues often had fig leaves. On the ones that didn't, the penis itself was so diminutive and innocuous that I doubted *it* could possibly be the same thing my father was hiding. Medical books were short on photographs of penises; even a realistic drawing was hard to come by. No classroom lectures on safe sex, with life-size models to fit with condoms. The boys saw films of oozing sores (but not one healthy member) in their health classes; we girls only got the female reproductive tract and pep talks on how wonderful it was to have a period.

When I was growing up, male movie stars and models provided no instruction either. Beautiful male bodies, of course, were not absent from popular cultural representation—particularly if you include athletes and movie stars as well as classical figures. Sexy male physiques— Paul Newman, William Holden, Yul Brynner—were available for fantasy even when I was growing up. But until very recently—outside of homoerotic photography and porn—a naked (usually hairless) male chest was the most one could expect to see, and that rarely. It wasn't until 1972, remember, that *Cosmopolitan* published that daring centerfold of Burt Reynolds, his penis hidden demurely behind his hands. "Equality at last!" Helen Gurley Brown declared.

By today's standards that centerfold certainly doesn't look like much; the fact that it was regarded as such a breakthrough is a good indication of how revolutionary the mere suggestion of unclothed penile regions was in those days. Now that I've become a student of such matters I realize that the Reynolds centerfold, as tame as it now seems, *did* represent something of a cultural turning point. It wasn't only (or even primarily) feminism that was responsible, though. As the politically oriented rebellions of the sixties gave way to the sex-and-lifestyle conceptions of liberation of the seventies, men's bodies began to be drawn into the ever-widening vortex of late-twentieth-century consumerism. Men could be encouraged to spend their money on fashion, hairstyles, jewelry, too! And what's more, there were plenty of previously ignored consumers out there—both male and female—who liked to look at male bodies.

Five years after the Reynolds centerfold made its appearance, Hollywood put its first hunk in (discreetly black) briefs on the screen (John Travolta, playing Tony Manero in *Saturday Night Fever*), and Calvin Klein, inspired by muscular yet sinewy gay male aesthetics, brought the beauty of men in tight jeans—and a bit later, clinging underwear— to a mass market. No naked penises, true. But a new willingness to visually foreground the sexuality of male hips and buttocks, and, ultimately, male genitals. The representational frontiers of the male body had been expanded; geographically, it now included a southern hemisphere. Consumer culture had discovered and begun to develop the untapped resources of the male body.

In the meantime, the penis was preparing to come out of the closet in less visual ways too. "Within cultural practice generally," philosopher Maxine Sheets-Johnstone has written, "a male's body is not anatomized nor is it ever made into an object of study in the same way as female bodies." Sheets-Johnstone overstates the case somewhat. There have been periods in Western culture—ancient Greece for one, the Victorian era for another—during which the sexual functioning of the male body seems to have been an object of intense scholarly or medical fascination. The advent of the "potency pill" Viagra in the late 1990s has made it an object of interest once more. But in general, Sheets-Johnstone is right. Male scientists and philosophers have created a nearly unbroken historical stream of tracts—philosophical, religious, scientific—on women's bodies and their distinctive maladies and excesses, almost all linked to our reproductive systems and sexual organs. But they have been remarkably good at forgetting that men have a sex.

In *The Second Sex*, Simone de Beauvoir explains why: "Woman has ovaries, a uterus; these peculiarities imprison her in her subjectivity, circumscribe her within the limits of her own nature. It is often said that she thinks with her glands. Man superbly ignores the fact that his own anatomy also includes glands, such as the testicles, and that they secrete hormones. He thinks of his body as a direct and normal connection with the world, which he believes he apprehends objectively, whereas he regards the body of woman as a hindrance, a prison, weighed down by everything peculiar to it."

Translation: To be a body with a sex is fine for girls—in fact, it's what we're supposed to be. But men are not supposed to be guided by the rhythms of bodily cycles, susceptible to hormonal tides. They are not supposed to be slaves to sexual moods and needs, to physical and emotional dependency. They are supposed to think objectively—to think like Man with a capital letter, discerner of Eternal Truth, the Universal Subject of History, Philosophy, Religion. They're not supposed to think with their penises!

When the emperor takes off his pants, however, it's hard to deny the fact that his body has a specific *sex* which (no matter how culturally "constructed") limits and influences him. So men just keep those pants on (metaphorically speaking), identifying completely with the products of their intellect and treating the penis (as cultural theorist Roger Horrocks puts it) "as an unfortunate by-product of evolution." A friend of mine tells a revealing story about a leading researcher on body image. This writer had written about every nook and cranny of men and women's bodies: pubic hair, chest hair, armpits, cleavages, women's genitals. But nothing, anywhere, about the penis. My friend asked him the reason for this absence. "Hmm, I hadn't really thought about it," the famous body researcher replied.

This willful forgetfulness is fragile, however. In every man's life are

memories of being led, as it were, by his penis. "[W]ho wins the argument with a hard-on? . . . When the prick stands up, the brains get
buried in the ground! When the prick stands up, the brains are as good
as dead!" That's Philip Roth, in his Alexander Portnoy persona. Portnoy has a particularly difficult time keeping his penis in line. So did St.
Augustine—who described himself as ruled (before his conversion) by
two laws, one of the heart, spirit, and intellect, the other "the law of
lust that is in my member." Augustine, like Portnoy, spent most of his
youth ridden with guilt over unwanted erections. But there are other
ways a penis can lead a man into indignity and humiliation besides
getting hard at inappropriate times, or when the conscience says no.
There's also the problem of getting soft when the *will* is intent on being hard, when being hard seems expected, required, when the desire
of another is waiting, judging, ready to be hurt, or angry, or resigned.

Men sometimes describe the penis as "having a mind of its own"—
and that mind is rarely seen as admirable in its mode of "reasoning."
Often, the penis is likened to a pathetic little child . . . or, on the other
hand, a conscienceless jungle beast. John Updike has said that men's
bodies feel "only partly theirs" because of the "demon of sorts . . . attached to their lower torsos, whose performance is erratic and whose
errands seem, at times, ridiculous. It is like having a (much) smaller
brother toward whom you feel both fond and impatient; if he is you, it
is you in curiously simplified and ignoble form." Alexander Portnoy is
less affectionate in his description; he's unzipping his pants stealthily
beside a sleeping shiksa on the bus from New York to New Jersey, and
". . . there it is again, up it pops again, as always swollen, bursting
with demands, like some idiot microcephalic making his parents' life a
misery with his simpleton's insatiable needs . . ."

A common source of shame among men is the fear that their sexual
impulses, written inescapably and unambiguously on the erect form of
the penis, will be unwanted. "Here," writes John Updike, "where
maleness gathers to a quintessence of itself, there can be no hiding; for
sheer nakedness, there is nothing like a hopeful phallus; its aggressive
shape is indivisible from its tender-skinned vulnerability." Some have
come to believe that their penises are ugly to look at. Nancy Friday, always ready to blame mothers, claims that it's women—"the sex who
thinks everything between the legs is dirty"—who have taught men
that the penis is disgusting. But the example she quotes from Tim

Allen's autobiography, describing his reaction to his father's penis (and sounding remarkably like Becky Stone's reports) suggests otherwise:

"My father would take me and my brothers to pee, and you're just dick tall, and your dad's is out. This whale of a penis would fly out, and you have a mushroom cap that two hands could barely pull out from your body. And your dad's penis would—thrummm! And you'd scream at this huge, hairy beast of an ugly—'Goddamn! Aw, God!' And we'd leave the bathroom and all go, 'Shit! Did you see that? Goddammit, it was all hairy, you know?' And we all prayed: 'I hope I never look like that!' "

Of course, at a certain point, having a "whale of a penis" becomes a not altogether unappealing thing. But there are still those associations, for many men, with the ugly, the instinctual, the "primitive" nature of the penis's existence. "You can think with your head all day, and maybe all night," Roger Horrocks writes, "but you must go to the toilet, and pull out your penis. You wake up early in the morning and it's hard. Think your way out of that one!" Yet the penis also reminds us, as one of my students aptly put it, "that [men] are just people, just some mothers' sons who once needed their weenies washed and probably will again." Is it any wonder, then, that it took Hollywood so many years before it allowed viewers their first brief glimpse of a naked penis? (It was Tom Berenger, playing a man in retreat from "civilization" in *At Play in the Fields of the Lord*.) After that, we occasionally got frontal nudity, but "full" is hardly how to describe it. "Flashing" is more accurate, as some man streaked across the screen, en route and with great dispatch: into the lake, scurrying to the bathroom. Come on, Mom; catch me if you can!

the return of the repressed

At the beginning of the nineties, things began to bubble and brew in unprecedented ways, bringing many of these elements up to the surface of American cultural life. 1991: Long Dong Silver—not only a porno character but a racially stereotyped one at that—referred to in extended (so to speak) discussion on the floor of the United States Senate. 1992: *The Crying Game*. First, one guy asks another to take his penis out of his pants to help him urinate, then we get a close-up of an actual penis—and on the body of what we thought was a woman! 1993: John Bobbitt's member is cut off by his wife (and chucked out the car window like a cigar butt), making real for thousands of men what had previously existed only in Philip Roth novels, therapy sessions, and bad nightmares.

It was as though the repressed had decided to erupt with a vengeance, straight from the collective unconscious, in all its most disturbing, marginal, anxiety-laden forms. "I've had enough of this closet. I'm going to throw open the door, and *really* make them squirm." And squirm we did. After Lorena made her point, the men of this nation (notwithstanding the fact that penis-severing is very, very rare—I personally haven't read of another besides Lorena's) became a country of Alexander Portnoys, nervously awaiting the moment when some woman would come at them with "the knife." Freud had told us all about it, hadn't he: the little boy whose organ is "so dear to him" will have it "taken away from him if he shows his interest in it too plainly." But that was supposed to be a fantasy! Although Freud did speculate, didn't he, that in "the human family's primeval period," castration actually was used as a punishment against naughty boys? He was right! And it's happening again!

Lorena was for some men what Mark Fuhrman would later become for some blacks—a symbol of the most heinous violence human beings can do to each other. One man wrote to *People* magazine that "being a male in America today is like being a Jew in Nazi Germany." Another wrote, in the *Los Angeles Times*, that "her abuse of him was so barbaric that the fact she was allegedly abused is hardly an issue." Rush Limbaugh and other conservative commentators hallucinated armies of hatchet-faced feminazis for whom castration was an exquisitely just revolutionary act, and Lorena a heroine for all battered women. (In fact, the few feminist intellectuals who cheered for Lorena and saw her action as a "wake-up call" to men included Katie Roiphe and Camille Paglia—not exactly the male-hating "victim-feminists" whom Rush had in mind.)

Our culture was disproportionately obsessional too about "the scene" in Neil Jordan's *The Crying Game*, in which archetypally femme-y singer Dil takes off a satin robe and reveals a penis beneath. In an earlier scene, Dil's ex-lover Jody, chiding his embarrassed captor Fergus into unzipping his pants to help him pee (his own hands are cuffed, since he's being held prisoner), had reminded Fergus that "it's just a piece of meat." Public responses to the movie, however, revealed that the penis is much more than that for us. "Don't give away the secret!" "Don't tell the surprise!" The degree of cultural collusion maintained about "*the* scene," in a society where leakage of information is a national disease, was truly remarkable. Not even the shower scene in *Psycho*—shocking at the time not only because of its violence but because it killed off the star so early in the film—was so carefully guarded. At times, the hysteria to conceal "the secret" of *The Crying Game* bordered on crazy absurdity, as when Roger Ebert furiously scolded Gene Siskel for revealing it on their "If We Picked the Oscars" show, even though film clips had just been shown in which Jaye Davidson, nominated for best supporting *actor*, appears as a woman.

Critics were dazzled. "In the middle of the London sequence, Jordan surprises us so thoroughly that we have to reconsider everything we have felt up to this point," wrote *New York* magazine critic David Denby. Wow. Not all of us, it's true, were quite so surprised, or felt we had to reconsider *everything*. Savvier viewers suspected Dil's biological sex from the start, on the basis of Jaye Davidson's thick neck and wrists, deep voice, and numerous other small but telling cues. Some continued to see Dil as an attractive woman even after her penis had been exposed. They agreed with Dil that the fact that she has a penis is just "details, baby, details" (as she notes drolly in another scene); penis or no penis, she's still a girl according to the cultural grammar of gender. But for most viewers, including Dil's hopeful lover Fergus (who promptly throws up at the sight of Dil's organ), the presence of that penis was definitive—the "absolute insignia of maleness," as psychoanalyst Robert Stoller has described it—and (unlike Fergus, whose stomach eventually settles down) we just couldn't get the sight of it on the "wrong body" out of our minds.

It may seem only natural to those who equate gender with biology that the presence of a penis would confirm that the body who has it is male. But not all sexual body parts scream out their gender as definitively as the penis does. Psychologists Suzanne Kessler and Wendy McKenna showed subjects two sketches: one with all the expected female sexual attributes—breasts, hips, long hair—but *with* a penis, and one without breasts, hips, or body hair but *with* a vagina. They found that the presence of the penis was the single most powerful, *the* defini-

tive cue for deciding which gender the figure was: 96 percent of their
subjects judged the figure with the penis as male *despite* breasts and
other female cues, while 33 percent of their subjects were able to ignore the vagina as a female cue in the other figure. *The Crying Game*
unnerved people by confronting them with a body that unsettled their
assumptions about gender—and raised the possibility of it all being or-
ganized very differently.

If Lorena Bobbitt stirred up castration anxiety, and *The Crying
Game* elicited what might be called gender anxiety, the Clarence
Thomas hearings thrust in our faces not only the prevalence of sexual
harassment but also a historical legacy of sexually tinged racism. Since
1991, we've nationally televised so many grotesque absurdist moments
involving sexual body parts—including news conferences on Bill Clinton's "distinctive" penile features—that it may take some straining to
remember how bizarre it was to hear the words "Long Dong Silver"
actually coming out of Orrin Hatch's prune mouth. He led up to it la-
boriously, strategically. "And she said 'He described pornography with
people engaging in oral sex.' Is *that* a black stereotype?" "No," replied
Thomas. Hatch: "People engaging in acts of sex with animals?"
Thomas: "No." Hatch: " 'Long Dong Silver.' Is *that* a black stereotype? Something like 'Long Dong Silver'?" When Thomas said yes,
Hatch performed the outraged innocent with gusto. "Well! I'm concerned! . . . This really bothers me!"

Hatch's shock was feigned for effect of course. Nonetheless, it was
an unprecedented moment, the prissy little senator from Utah—
Utah!—forced to say "dong" in order to achieve his political goals.
The senators didn't even like to say the word "penis" let alone refer to
a gigantic black one. While they thought nothing of dragging Anita
Hill through hours of humiliating testimony about pubic hairs on
Coke cans and various genres of pornography, they could bring themselves to say the p-word—and with obvious discomfort—only when
forced to, quoting from Anita Hill's testimony. Hill's own lawyers too,
arguing that "the decorum of the Senate favored delicacy," avoided the
term "penis." Their preferred term was "private parts." Even today,
we sometimes have trouble with the words for male genitals. ("Stop
cancer *down there*," coyly states the July 1997 cover of *Men's Fitness*.)
We've become far less delicate, however, about public intrusions into

private sexual behavior. We may still employ the euphemisms, but not even—no, let me rephrase that, *especially* not—the President of the United States is actually permitted to have "parts" that are even remotely considered private.

The senator's discomfort was nothing compared with Thomas's. "Yes," Thomas had replied through clenched teeth in response to Hatch's deliberately leading question, "the size of sexual organs would be something . . ." It was an awful moment, no matter whose side one was on. Thomas could barely get the words out, groping to make the stereotype clear while at the same time distancing himself from it; his use of the subjunctive mood (as though he were describing a possible universe), the vagueness of "the size": unconscious protection against his own contamination by the image he was exploiting. Thomas was hoping that the charge that he'd been tarred by racial stereotypes might win him his seat on the Supreme Court—and it did. But he also knew that racist imagery, once released from the collective unconscious, is apt to run amok. And so it did. Thomas won, but a sympathetic *People* magazine story which appeared later that week, showing Thomas piously reading the bible on the couch with his wife, Virginia, made Thomas's crotch the visual focus of every photograph.

reading bodies

In 1998, *People*'s photos of Thomas may not be as striking—or as offensive—as they seemed to me in 1991. For sometime during the last seven years, we seem to have entered an unabashed Reign of the Ram. Fashion ads today routinely feature men in underwear who are amazingly endowed ("He looks like he has *two* penises in there!" a student of mine said of one), their poses directing the viewer's attention to their crotches. Some ads show men cupping their genitals; in others, the penis seems to be fully erect. *Playgirl*, clearly doing double duty for both gay and straight readers, has become a cornucopia of humongous male organs of every race, class, and age. Rap musicians (and, aping them, Madonna) have made crotch grabbing a common cultural code for sexual power and virility. Surgical phalloplasty is big business today—and, by the way, has provided John Bobbitt with a new, happier

ending to his saga. Bobbit has not only been re-membered but enhanced ("It's like a beer can," he boasted after the operation). So will
someone send a thank-you card to Lorena already?

This is not the first time in history that penises have been culturally
visible. Modernity has been especially squeamish, it seems, about the
male body. Until the creation of trousers at the beginning of the eighteenth century, men's pants were more like what we call "tights,"
through which the shape of the genitals could often be clearly discerned. However, as Anne Hollander points out in *Sex and Suits*, the
simulation or suggestion of male nakedness had a very different significance then than it does today. In 1998, we look at the frontal bulges
of fashion models clad in clinging jersey briefs and think: "Sex." From
the early Renaissance to the eighteenth century, it was just the opposite. Revealing the basic outline of male legs and genitals while covering those parts of a woman's body with voluminous materials, drapes,
hoops, and the like meant the male body was a more utilitarian, "authentic," no-nonsense, *truthful* body, while the woman's body was an
object of artifice, fantasy, and illusion, her clothing designed to stimulate the sexual imagination rather than represent her anatomy accurately.

The example shows that exploring attitudes toward clothing, nakedness, and masculinity means recognizing their historical and cultural
variability. It also suggests that we need to think about the body not
only as a physical entity—which it assuredly is—but also as a cultural
form that carries *meaning* with it. The notion that bodies are just material things, collections of instincts, mechanical processes put in place
by God the watchmaker, ticking away in pretty much the same way
from culture to culture, era to era, goes back to the philosopher
Descartes. Darwinian science profoundly altered the Cartesian picture
by showing that the body's mechanisms aren't timeless and unchang-
ing, but have evolved dramatically over time. What's still missing from
this picture, though—and what was ultimately supplied in the twentieth century—is the recognition that when we look at bodies (including
our own in the mirror), we don't just see biological nature at work,
but values and ideals, differences and similarities that *culture* has
"written," so to speak, on those bodies.

What this means is that the body doesn't carry only DNA, it also
carries human history with it. Biologically, a penis is a penis (more or

less). But as Hollander's research shows, a seventeenth-century spectator would have seen something very different than we do when confronted with the sight of a man in genital-revealing tights. Another
example of how history and culture affect the way we see bodies is
provided by the Jockey ad above. The two parts to the ad are paral-
lel—five good-looking men (doctors, surgeons, and physical therapists)
and five good-looking women (riders, ranchers, and cowgirls) with

their pants around their ankles, revealing their Jockeys underneath. Some features of these ads are pretty obvious. The open expressions on the women's faces versus the mostly serious, semi-challenging male stares; the men's "feet planted firm" posture versus the women's flirtatiously bent knees—these are standard stuff in the depiction of men and women's bodies. As for what's "new"—Jockey clearly means to tweak the cultural stereotypes of doctors and cowboys with its unexpectedly muscular, macho docs and curvy, feminine ranchers. And to convey a message of equality too, by having the women and men in equal states of undress.

The thing is, they're *not* in equal states of undress—because pants-around-the-ankles convey something different on the bodies of the men than they do on the bodies of the women. The men's genitals in the Jockey ad are much more visible than the women's. Actually, the women, by contemporary standards, are hardly undressed at all. Yet somehow, they *seem* more exposed than the men. It's those pants around the ankles. Mentally remove them—or just cover them up with your hands—and you get a very different feeling from the ad. With the men, on the other hand, the "dropped trou"—the expression seems appropriate for them, but not for the girls—barely signify at all. Cover them up or leave them in, the guys' bodies project much the same confident, slightly challenging machismo either way. Their bodies are very sexy. But they do not seem stripped or exposed.

Partly, this is because we read the bodies in the Jockey ad against the backdrop of depictions of male and female bodies that we are used to, and until recently, the conventions governing the two sexes have been very different. In countless movies, we've grown used to seeing women take off their clothes for sex, or to display themselves erotically, or to be unsuspectingly spied on. The act of getting undressed is an act of uncovering, exposing; a secret sexual self is revealed. In contrast, men, until very recently, would only be shown undressing with some "utilitarian" fiction written into the scene—changing from business suits into sweats, putting on athletic uniforms, and so on. Anne Hollander would probably argue that all of this has to do with the lingering of those old ideas about the utilitarian form of the male body. And when you examine the accessories that have been included in the Jockey ads, it seems that there's something to that. The docs seem ready to get to work. They've got their stethoscopes around their necks, their surgical masks ready to tie on; all they need to do is pull up those pants, whip out that knife, and start bossing the nurses around. The cowgirls, in contrast, are ill equipped to round up anything beyond male winks. (Those hats are cute, though.)

There's a more unnerving piece to this too. With women, pants-around-the-ankles subtly recalls movie scenes of rape (think Jodie Foster on that pinball machine in *The Accused*) and the discovered bodies of female murder victims. Such images are not part of a feminist's paranoid delusion, but belong to a common lexicon of cultural motifs that we are almost all familiar with and are very difficult (if not impossible) to erase from the unconscious. The fact that the models in the Jockey ad are *not* coming on like sexual temptresses but have bright, sunny smiles only enhances the lascivious juxtaposition between their own definition of themselves (as active, happy girls) and the perspective of the viewer, in whose eyes they seem vulnerable, unknowingly exposed. You feel, in some subliminal corner of the brain, that while the men have "dropped trou" themselves someone has pulled those girls' pants down (or asked them to pull them down, or is peeping around the corner). Yes, I know the *actual* models don't feel that way at all . . . and would probably have a good hoot if they read this too. I'm not talking about how the models *feel* but about elements suggested in the representation they posed for.

You may not see the same things in this ad that I do. Representations of the body have a history, but so too do viewers, and they bring that history—both personal and cultural—to their perception and interpretation. Different viewers may see different things. In pointing to certain elements in ads, or movies, or fashion, I'm not ignoring the differences in how people may see things, but deliberately trying to direct your attention to what I see as significant. I'm making an argument, and you may or may not find it convincing. You might think—as my students sometimes do—that I'm "making too much" of certain elements. Or your own background, values, "ways of seeing" may enable you to discern things that I do not. Or it may be that even as I offer my ideas, the cultural context has already begun to shift in ways that answer my interpretation or make it obsolete. Cultural interpretation is an ongoing, always incomplete process, and no one gets the final word.

Hollander's insights, for example, while useful in interpreting some aspects of the Jockey ad, ultimately come up against the fact that we *do* live in the second half of the twentieth century, and despite the utilitarian accessorizing, the doctors' discernible genitals, no less than the girls' body parts, *are* the stuff of sex. With all the naked bodies around nowadays, this sensibility may ultimately reach its historical limit too, as we get jaded and bored with all the (mostly plastic) flesh around. But for now, as a general rule, the more a piece of clothing outlines and reveals what's underneath, the "sexier" it is to our cultural eyes. For the execs at Mattel, the Ken doll's little bump did not make him more "authentic," it made him pornographic.

The kind of sex the docs project is different, though, than the sex the cowgirls project. Nowadays, men—the Chippendales, for example, and men in ads as well—*do* strip for erotic display. But when they do so, they tend to present their bodies aggressively and so rarely seem truly exposed. Being undressed does not necessarily mean being naked. Most men in underwear ads today . . . Well, you wouldn't want to meet one coming round a dark corner. When, in the 1996 French film *Ridicule*, an aristocrat took his penis out of his pants and pissed in the lap of another man, he seemed more to be whipping out a weapon than a body part. And what can a viewer's eyes do but admire the bulging bundles and gleaming muscles of the Chippendales and their ilk? Their bodies are a kind of natural armor. Being "stripped," it seems—much more than merely being undressed—has been anathema to masculinity. (Significantly, the plan to put a permanent swimsuit on Ken was abandoned when Charlotte Johnson, Barbie's clothing designer, pointed out that the "clinging" swimsuit would only make Ken's "hidden" parts more enticing. Little girls, she argued, would undoubtedly scrabble at that painted swimsuit and scratch it off anyway—an invasion of Ken's privacy that seemed even more loathsome to the male execs. Better that Ken be undressed to begin with than have little girls strip him of his dignity.)

reign of the ram or emergence of everyman?

It sometimes seems as though popular culture has gone directly from near-censorship to blatant sexual fetishization—even idolatry—of the male organ. The wildly popular television comedy series *Ally McBeal* devoted two episodes to Ally's temptation by (and ultimate affair with) an artist's model who is so fabulously endowed that Ally's foxy roommate Renee, doing a clay sculpture of him in class, has to ask for more clay. We don't get to see the original (only the bulging eyes and open mouths of Ally and Renee), but at one point Renee's clay simulacrum falls to the floor (gravity has screwed up her attempts to do the statue "to scale") looking pretty darn lifelike. After class, at a restaurant, Ally and Renee wave rubbery sausages about, as they discuss whether the model's member was an implant or natural ("*super*natural," suggests Renee). When Ally accepts a date with the guy, her female colleagues—who have spent hours standing around discussing the size of this guy's penis—snigger "Don't get hurt!" They don't mean emotionally.

Sexual politics as well as consumerism have clearly played a role here. Between 1991 and Long Dong Silver and 1998 and Long John Silver (as one of the women refers to the model) watercooler talk about big penises has apparently become not only "correct" but obligatory for demonstrating that the heroines of your show are not *those* kind of feminists. You know, the Anita Hill kind. Any tendencies in that direction are always countered on *Ally McBeal*—not with an "argument" (this is postmodern television, not some dreary old *L.A. Law*), but with irony. Ally, for example, feels guilty for being attracted to the model with the shlong. Less repressed Renee (she's a black woman, after all) points out that men have been no less superficially magnetized by big breasts (including her own "golden, lofty globes"), so why not go for it. "But we're women," Ally sputters, "we're different . . . we have double standards to live up to." The show itself is dedicated to blasting those double standards to pieces—when the mood strikes it—and keeping gender very, very traditional when *that* suits it. In another scene in the same show, Ally—having successfully defended a young man who broke another guy's jaw for insulting his girlfriend—admits while strolling with the artist's model that she too would "want my date to rip off the head" of any guy who insulted her. As I've said, Ally is a postmodern show. It gets to have it any way it likes.

So, although the women on the show are post-Hill feminists, the men on the show are post-Thomas guys too—at least when it comes to boasts about their own penis size. "Does size really matter?" Billy

(Ally's ex, now married to another female lawyer at the firm) asks the giggling women, as they stand around discussing the model's "trunk." As is the case for many ordinary guys in the real world, the super-endowed model has bred insecurity—*not* identification—in Billy. The girls (these are not the kind of girls to be offended by being called that) assure Billy that finger size is more than enough for them. But they're lying, as their giggles demonstrate to Billy. Later, in bed, he's unable to maintain an erection with his wife—which touches her, since until then she'd always been the insecure one in the relationship. But in the meantime, across town, Ally is having primally satisfying sex with Long John Silver, interspersed with scenes of her colleagues (male and female alike) watching a boxing match on television, celebrating their earlier legal vindication of man's "warrior nature," smoking big stinky cigars, lustily urging their favorite to take the other guy out.

Depending on your perspective, the contradictions of *Ally McBeal* are either wonderfully true to the complexity of contemporary life or a shameless, cynical media potpourri, with Teflon politics to which nothing sticks. In the case of the big penis show, however, the show's contradictions accurately (if unquestioningly) reflect our current ambivalence about masculinity, itself reflected in the different guises with which the penis has come out of the closet in our culture. On the one hand, there are the underwear models with jaws and balls of steel—the phallic mythology of Superman masculinity, still vital if not entirely intact in this culture. Think, for example, of our definition of "erectile dysfunction," a term which has entered everyday conversation with the marketing of the miracle drug Viagra. With so much controversy about Viagra (a development I'll talk about more in another chapter), it's becoming somewhat politically incorrect to use the old term "impotence." That's all to the good. But instead, we say "dysfunction" and then define it as the inability to achieve an erection that is adequate for "satisfactory sexual performance"! Not pleasure. Not feeling. *Performance.* Eighty-five-year-old men are having Viagra heart attacks trying to keep those power tools running.

On the other hand, there's the less than masterful reality of Everyman, also coming out of the closet. In the second half of the 1990s, we've seen men depicted learning what it's like—in the charming and successful *The Full Monty*—to be on public display *without* that gleaming phallic armor. We've heard male insecurities about penis size verbalized on national television. (*Seinfeld* was the first to break that ground, with Costanza's frenetic efforts—on one of the show's most hilarious episodes—to make sure that Jerry's date for a beach weekend, who had accidentally seen his unclothed penis as he came out of the water, will know that he's not *really* as small as he looked. "Shrinkage!" he keeps insisting.) We've even seen a penis caught in a zipper, in the over-the-top comedy *There's Something About Mary*. Now what could be less masterful than that?

Superman and Everyman, however, don't live together in real life as cozily as they do on *Ally McBeal*. For Superman haunts Everyman, threatens his undoing, as Paul Thomas Anderson's *Boogie Nights*—my own favorite coming-out film—dramatizes. The film is the story of the rise and fall of a mythically endowed (or so we are led to fantasize throughout the movie, from the amazed eyes of those who gaze on him) young porn star, Dirk Diggler. Diggler (a stage name) is played by Mark Wahlberg, who used to go by a stage name of his own—Marky Mark, who in his former cultural incarnation as rapper and underwear model did his own bit in bringing the penis out for popular consumption. Burt Reynolds, that centerfold innovator, plays director and porn-film producer Jack Horner. (This film is a field day for what poststructuralists call "intertextuality.")

"There's something wonderful waiting to get out of those jeans," Horner tells young Dirk. And as long as Diggler is revered as the biggest, hottest kid on the block, he does just fine. But he loses his grip in the face of competition, and begins to have trouble getting it up. This dilemma is no scriptwriter's contrivance. With the "money shot" (ejaculation) the culminating moment of virtually all porn films, the pressure for erection on command is constant for the male porn star. Susan Faludi reports a conversation in which an aspiring actor asked an experienced actress for pointers in the business. "Just get it hard," she told him. But this, of course, is more easily said than done, and frequently the actor—with the whole crew tapping their feet—must "wait for wood." Often, the actors fail and "break down on the set" (lose their erections) and lose the gig. Thus the world of pornography, which (as Ira Levine has put it) dramatizes "the triumph of the dick," at the same time tyrannizes the actors who are expected to enact that triumph.

On the surface, a commercial underground where men pray for

"wood" and lose their jobs if they cannot achieve erection on command is far removed from the lives of most men. On a deeper level, however (and as *Boogie Nights* illustrates), the world of the porn actor is simply the most literalized embodiment—and a perfect metaphor—for a masculinity that demands constant performance from men. When Wahlberg first heard about the script to *Boogie Nights*, it unnerved him, reminded him of his own life, and "all the stuff I wanted to get away from." Wahlberg was a male model, of course, and one famous for his endowments. But a man doesn't have to have made his fortune through his member in order to identify with Dirk Diggler's predicament. (The man with whom I saw the movie—a college professor—told me his palms were sweating throughout it, anticipating the inevitable failure.)

Even before Diggler takes up a career that depends on it, his sense of self is constellated around his penis; he pumps up his ego by looking in the mirror and—like a coach mesmerizing his team before a game—intoning mantras about his superior gifts. That works well, so long as he believes it. But unlike a real power tool, the motor of male worth can't simply be switched on and off. Unwavering, unbending power may be intoxicating as long as the fortress holds; but as soon as it cracks, the whole structure falls to pieces. And it's bound to crack; it's just too much to expect from a human penis—or a human being. In the final shot of the movie, we see Diggler's fabled organ itself. It's a prosthesis, actually. (Some men thought that was a cheat—until I reminded them of breast implants.) But prosthesis or not and despite its dimensions, Diggler's penis is no masterful tool. It points downward, weighted with expectation, with shame, looking tired and used.

I don't have a penis, and I don't intend to speak in this book as an expert on those who do. My own experience with the male body is as a woman, a perspective that will be kept at the forefront rather than obscured. This has advantages as well as drawbacks. There are things I can never know by virtue of the fact that I don't have my own "ignoble little brother." But by the same token, I have no need to disown, deny, or despise myself for his mischief. A certain empathy for the special predicaments of the male body, ironically, may come easier to someone who *doesn't* have one.

At the same time, it may be that this most male of bodily sites—the penis—holds the most promise for a deeper identification between men and women. Dirk Diggler's predicament—shame, exhaustion with cultural expectations, the failure of the body to live up to those expectations—these are all experiences a woman can relate to. I have been amazed at how much unexpected kinship I've felt with men while writing this book, and how many old myths I have been led to revisit and revise. In the process, I've come to see that the pop psychologists are dead wrong. For all our differences, so entertaining and lucrative to emphasize nowadays, men and women do *not* come from different planets. One of my goals in writing this book is to demonstrate that.

hard and soft

the sex which is *not* "one"

French feminist Luce Irigaray has described woman as "this sex which is not one"—a phrase meant to be understood with the emphasis on "one." Women's sexuality, as Irigaray celebrates it—in poetic and sometimes obscure language which some people adore and others can't bear—is not singular but multiple. It has no fixed location in the body (despite what sexologists and some early feminists have said about the clitoris as essential to female sexual orgasm) but is capable of being experienced and expressed all over the body. Touch. Smell. Imagination. Looking to the "little penis"—the clitoris—as the sole route to orgasm is rather phallocentric, by Irigaray's lights. Rather, "woman has sex organs just about everywhere," Irigaray writes. Men, it is implied by contrast, have only *one*. By that, she seems to mean that men *are* their penises, sexually speaking.

In many ways, Irigaray is right. We live in a culture that encourages men to think of themselves as their penises, a culture that still conflates male sexuality with something we call "potency" and that gives men little encouragement to explore the rest of their bodies. Think, for example, of how many advertisements depict women stroking their necks, their faces, their legs, lost in sensual reverie, taking pleasure in touching themselves—all over. In such ads, women are continually given permission to relate to their full bodies, even to become immersed in them. Images of men in similar poses are very, very rare (I've been searching them out for years) and just about *never* employed by manufacturers and designers who are after a predominantly straight male market. Touching oneself languidly, lost in the sensual pleasure of the body, is too feminine, too "soft," for a real man. Crotch grabbing—that's another matter. That's tough, that's cool, that's putting it in their face.

Irigaray has a point to the extent that men see themselves as "one" with these cultural messages. She's right when it comes to those crotch-grabbing moments in rap, or the phallocentricities of a writer like Norman Mailer, or the pitches of cosmetic surgeons urging men to add inches and thus manhood to their members. She's certainly right about the rapist who uses his penis as a weapon, or even the husband whose only sexual move is in and out. She's nailed those men who equate sex with penetration, like the guys in Ken Smith's *Chasing Amy* (one says he can understand "fags," because they at least are "dicking" each other, but can't imagine how lesbians can have "real sex" without a strap-on).

The cultural equation of penis = male, as I discussed earlier, is what makes the penis-revealing scene of Neil Jordan's *The Crying Game* the pivotal moment in the film. For most people, it revealed Dil's "true" sex. All sorts of powerful feminine attributes of Dil's—more classically feminine than the film's "real" woman, the macho terrorist Jude (Miranda Richardson)—were simply mentally erased by many viewers once that penis was revealed. Many transgendered individuals too have viewed getting rid of or acquiring a penis as the fulcrum of their search for a body that will fit their gender identity. Many experience their genitals as "wrong"; Richard Raskind, who was to become tennis player Renee Richards, describes years of painful binding, tucking, taping, and tying in an effort to get his penis out of sight.

But just *where* does that sense of having the wrong genitals come from? It makes sense to me that it is produced, at least in part, by the culturally available framework for imagining the relation between gender identity and body in terms of genital fit or mismatch. If having a gender identity is tied to having a certain genital morphology, then of course one's morphology is going to be experienced either as corresponding to or in conflict with who one "really is."

"THE PROBLEM: NO PENIS" reads a slide presented by phalloplasty specialist Dr. Donald Laub during a talk at the first annual Female to Male Transsexual conference. Well, if "no penis" is *defined* as the problem, of course having one will be the solution. I'm sure Dr. Laub views himself as echoing sentiments and serving needs that are already there in his audience. And certainly, it's true that many transsexuals tell the story of their lives as the struggle of a very clear and singular inner identity to find its appropriate bodily form—the form designed to most perfectly enable their true nature to express itself. "There was to the presence of the penis something positive, thrusting, and muscular," reflects travel writer Jan (formerly James) Morris. "My body was then made to push and initiate, and it is now made to yield and accept . . ."

Morris speaks as though God or nature made the bodily roles that Morris describes here. But which body parts are "initiating" and which "yielding" does not simply follow from the facts of anatomy or intercourse, as we know from our many cultural representations of aggressive, biting, tearing vaginas. The way we experience our bodies is powerfully affected by the cultural metaphors that are available to us. I've been struck, reading more recent personal accounts, by how many transgendered individuals do *not* feel they need to alter their genitals in order to reconcile identity and body, or who do not view altering their genitals as the essence of that quest. (The new term "transgendered," which refers to all people who do not live under the "male" or "female" identity they were assigned at birth—whether or not they have had sex-change surgery—reflects this reality, which was obscured so long as Renee Richards and Jan Morris were our sole cultural paradigm.) Philosopher Henry Rubin writes:

> In general, I want to be unambiguously recognized as an authentic man even though I was not born with a male body. When addressed by my new name or the pronoun "he" I want the speaker to mean what she says. I do not want people to think of me as a woman first and then have to translate it in their heads, the way that you do when you learn a new language. I want to respond naturally to her address, not to look around to see if she meant me. I want to live manhood with the same authenticity as a man born with a male body. I want this so much more than I want a penis.

For Rubin, feeling like an "authentic man" is more a matter of having a certain kind of presence and reception in the world than a matter of having the right genitals. Rubin recognizes that people may respond in the desired way only to those whom they believe have the right genitals—a penis—but that recognition is a far cry from seeing the possession of a penis as necessary to reflect the inner "truth" of one's identity. Other transgendered individuals don't even experience the desire for gender authenticity that Rubin speaks of, and are skeptical of the very notion itself. They complain of having to fit their experience into a framework that insists that identity must be either male *or* female and that doesn't even provide grammatical forms to accurately depict their plural histories. (I've been struggling with that difficulty, as I try to figure out when to write "he" and when to write "she.")

The message I take from these new accounts is not that there is no biological component to sexual identity. There's plenty of evidence to suggest that there is, including the experiences of some of those children who've had malformed penises removed at birth, who—although they knew nothing about their operations—grew up feeling intense discomfort at being treated like girls (like "a trapped animal," as one boy put it; he later had a mastectomy and phalloplasty, and became "male" again). But whatever the input of biology, culture is always present, providing language and categories to help interpret that input. It's necessary to pay attention to both biology *and* culture—and their interaction—when thinking about the male body (or any body).

This is something that's rather hard to do, since it's not the way knowledge has been encouraged to develop in our culture. Instead, we conduct most of our research, conversation—and on a more basic level, our *thinking*—within academic disciplines, themselves divided into "the sciences" and "the humanities," each of which has been put in charge of certain parts of the world, certain kinds of inquiry, while regarding everything outside those borders as someone else's turf. Within our own borders we learn to do things in particular ways, we develop special standards of "rigor," we are trained to see certain things as more important than others. All of that's fine, and absolutely necessary for the acquisition of certain kinds of knowledge. But not all. There comes the time when things need to be put together into a bigger picture, a larger context, when connections need to be made if we are to have anything except a fragmented picture of the world.

At that point, the humanities and the sciences often find that they are ill equipped to talk to each other, let alone integrate their specialized research into something more whole. For the other side of the coin of our learning to see things in a certain way is the development of characteristic blind spots, which we bring with us even when we attempt more interdisciplinary work. Philosophers and literary theorists, for example, are now hot for "the body." But they often forget that the body is made of flesh and blood and has an evolutionary history. So Thomas Laquer, arguing that the notion of "two sexes" is entirely a cultural, historical creation, ignores the larger evolutionary picture and potentially instructive comparisons with other animals, including our own ancestors. We can't talk about other animals, he says, without interpreting their behavior on the basis of our expectations that we will find two sexes. So the evolutionary literature on sexual dimorphism is dispensed with as little more than a piece of ideology.

Scientists, on their part, ignore how extensively symbolic constructs feed into human history, influencing the evolution of flesh and blood. They are sometimes scornful when humanists talk about racial, sexual, and other cultural biases that can influence research. They often jump to the (ridiculous and inaccurate) conclusion that anyone who is interested in social influences on science (or on forms that scientists have been taught to think of in purely physiological terms—like the body) is denying that there is a "real world" out there.

Well, I am one humanist (and there are plenty of others too) who firmly believes in the reality of the world. But I would insist (again, with many others) that reality doesn't just depend on the laws of physics or the structure of DNA, but also on cultural images and ideology. A person's genetic inheritance may be a fact of nature. But that inheritance will set him up for a struggle with the social and sexual identities assigned to him only if those categories are too rigid to accommodate his experience. Some cultures have greater diversity among gender categories than we do. The *travestis* of Portugal are male prostitutes who adopt feminine names, clothing, and hairstyles, take female hormones and inject themselves with silicon in order to create breasts, wide hips, and large buttocks; yet they do not identify themselves as women. The *Khanith* ("effeminate") of Muslim Oman dress like women, perform tasks assigned to women, and are socially accepted as such. Similar roles exist or have existed in India, in the Philippines, and among North American Indian tribes. As an example from the extreme other end of the spectrum of cultural tolerance—and a pretty good indication of European attitudes until fairly recently—the Inquisition revoked Joan of Arc's life sentence and burned her at the stake when she persisted in dressing in men's clothing.

It takes two to tango here. If the available categories of social identity are flexible, fewer people will feel themselves—whatever their biological dispositions—in conflict with them. Part of what seems to be going on nowadays is that some people are trying to reconstruct the categories as well as their bodies. More profoundly, as people alter, expand, and experiment with—surgically and otherwise—the bodily forms that constitute our repertoire of sexual possibilities (or, for that matter, racial possibilities), the categories *inevitably* will become inadequate. Spend some time loitering around the halls of a middle-class high school, watching young people walk by. It's no longer as easy as it once was to figure out who is "gay" and who is "straight," who is "black" and who is "white." I put these categories in quotes to emphasize how socially mutable they are and the fact that people's realities were *never* as simple as we imagined them to be from certain bodily signs. Nowadays, the codes are getting even less reliable than they once were, as young people "mix it up"—genetically, sexually, stylistically.

In the case of race, the old dualism of black/white is being shattered not only by genetic mixing (which we've always had, but which over the generations has become more and more intermingled) but by a new generation of young people (like golfer Tiger Woods) who truly and honestly *don't* think of themselves as one race or another and are angry when forced to declare themselves in that way. Many teens and twenty-somethings feel the same way about sexual orientation. When Anne Heche came out as Ellen DeGeneres's lover, she did *not* describe herself as having discovered she was "really" a lesbian or even bisexual; she insisted that she simply fell in love with a particular person, who happened to be a woman. Contemporary fashions—in clothing, hair, walking, talking, dancing—offer a pastiche of sexual styles, with the emphasis on postmodern cool rather than "masculine" or "feminine" looks. I'm a pretty good "reader" of bodies, and I'm pretty literate too in the old bodily codes by which gays, lesbians, and straights used to announce their identities to each other. Nowadays, though, I'm frequently flummoxed.

This mixing-it-up is what contemporary theorists are talking about when they use the term "queer" to cover a whole range of sexual styles replacing the old dualistic categories of sexual orientation and gender "identity" which forced us to declare ourselves gay or straight, masculine or feminine, male or female. Those "queer" tides of our culture have affected ideas about masculinity. But I don't want to lose sight of possible biological counterinfluences, or powerful *cultural* countertides—older, stronger, resistant to change. It's my assessment that they are still bossing most of us around and need to be taken to task. There are significant generational differences—but race and class differences too—in how defined people feel by particular identities. Some people live in communities (the rich and famous, for example, or—to bring it a bit closer to earth—kids who go to "progressive" schools) that are more freewheeling or experimental than others. But we don't yet live in queer nation, perhaps never will.

That's where Irigaray's point has validity—at the level of old images and ideologies which, at times, seem to be remarkably intact in their definitions of manhood, and which still exercise a powerful effect on men's experience of their bodies. Want just one barometer of that effect? Look at Viagra—not the drug itself, but the way men (both users and doctors) talk about it: Performance, performance, performance. I haven't yet read one account in the newspapers or magazines in which a man talks about any increase in pleasure, either psychological or physical—beyond overwhelming relief, perhaps renewed pride. It's harder. It's firmer. It can go all night. New York urologist Steven Lamm describes one fifty-two-year-old patient (whose anxiety seems not at all unique, based on what I've read) who wanted Viagra as "insurance" in his relationship with a twenty-four-year-old woman. The man inquired, at the same time, about a drug to deal with his baldness.

I mention the baldness because I find it highly instructive. Although critics of the double standard in health care for men and women rightly point to the unfairness of insurance companies paying for Viagra but not birth control for women, a more exact analogy is probably not with birth control but with diet pills. Perhaps some men use Viagra to be able to conceive; but in all the newspaper and magazine stories I've read, that motivation hasn't been mentioned *once*. Viagra for men, like diet pills and cosmetic surgery for women, is not about restoring reproductive function—or even, to hear men talk, about restoring sexual pleasure—but about meeting and keeping up with the cultural standards and expectations of masculinity and femininity. The pressure is enormous on both sexes. But just as women try to become like the skinny models but don't really feel very happy about starving themselves, most men are not fully "one" with the cultural messages that tell them their power resides in their pants. Even those cultural messages are themselves not really unified, but bark out contradictory orders to men all the time. (Go for it, be a man, be a wild thing. But remember that no means no.) Ideas about the penis and masculinity, too, are hardly "one." There are ethnic and racial ideologies that attribute special characteristics to the "black penis" and the "Jewish penis." There are the playful, ironic sensibilities that endow the penis in gay male culture with the divided personality of sexual icon and object of parody. There's the mythic phallus, the cultural symbol of masculinity. And then there are penises of flesh and blood— clearly creatures of variety, not unity. Imagine a penis, social theorist Charles Bernheimer suggests. Now ask yourself what it looks like, and you'll find that many possibilities come into play. "Is it large or small, erect or flaccid, circumcised or not? . . . Is the skin white, black, yellow, or some in-between shade? Does your picture include the testicles? . . . Is there any suggestion of sperm? . . . What do you fantasize doing with it? How does it feel? smell?" And so on.

No, "one" is not the metaphor I'd use to describe the world of men.

tender buds, proud members, and metaphorical dildoes

No other part of the body is so visibly and overtly mercurial as the penis, capable of such dramatic transformation from passivity to alertness. No wonder many cultures have worshipped the phallus as a magical being. The word "fascinate" has its origin in the Latin word *fascinum*, which meant "witchcraft" and derived from the phallic god Fascinus, worshipped by Romans, who sometimes wore an image of an erect penis around the neck as an amulet or hung one on the walls of their houses. It was believed that the image had magic powers, could ward off the evil eye. "In time," Diane Ackerman writes, "anything worth appreciation and study, anything potent and magical, anything as truly terrific as a penis, was called fascinating." Some Renaissance churches claimed to have part of Christ's penis—his circumcised foreskin—as a holy relic, to which women prayed in order to help them conceive.

We don't *quite* regard the penis as a magical being, but we still find fascination in its mercurial nature. Irigaray herself notes, in passing, that "the passage from erection to detumescence does pose some problems" for her thesis about men being "one." I'd say that's putting it mildly. "Hard" and "soft" are two dramatically different physiological states that have been endowed with even more dramatic—and varied—significance by culture.

Nonerect, the penis has a unique ability to suggest vulnerability, fragility, a sleepy sweetness. It's not just soft; it's *really* soft. It lolls, can be gently played with, cuddled. Few other parts of the body, especially in our implanted culture, are quite like that. Even breasts, nowadays, are likely to be harder than a soft penis. In literature, tender descriptions of the penis are usually evoked when it is soft. The most famous is offered by D. H. Lawrence through the persona of Connie Chatterley, who murmurs to Mellors's soft penis as though it were her infant baby, even a fetus.

> "And now he's tiny, and soft like a little bud of life!" she said, taking the soft small penis in her hand. "Isn't he somehow lovely! so on his own, so strange! And *so* innocent! And he comes so far into me! You must *never* insult him, you know. He's mine too. He's not only yours. He's mine! And so lovely and innocent!" And she held the penis soft in her hand.

Say what you will about Lawrence in his phallic postures; he's truly captured a woman's moment here.

All animals, of course, are made of mostly soft stuff, requiring various kinds of protection, from horns to helmets, to help us get by. Human flesh is particularly vulnerable, but the soft penis seems especially so, not, I think, because (like the testicles) it *is* more easily hurt than other parts of the body, but by virtue of contrast with its erect state. No other body part offers that contrast. Unfortunately, the relation between the hard and soft penis often determines whether the soft penis will be cherished like a sleeping baby or derided as a flaccid piece of failure. The vulnerability of Mellors's soft penis touches Connie Chatterley, but only after she has known him in a more commanding mode; transfigured by satisfying sex, she looks with wonder at "the tender frailty of that which had been the power." But fifty pages earlier, after unsatisfying sex, Mellors's body appears a "foolish, impudent, imperfect thing," the "wilting of the poor, insignificant, moist little penis" of her lover to be "ridiculous" and "farcical." Whether he's a tender bud, full of promise, or a sad, wilted bloom, the full flower of manhood sets the standard.

The erect penis is often endowed with a tumescent *consciousness* that is bold, unafraid, at the ready. Gay art and literature and both straight and gay pornography are throbbing with such descriptions, and so nowadays—as I recently discovered—is the romance novel.

I wanted to survey the sexual metaphors in these novels. So I asked a woman perusing the romance shelf of my local bookstore for some recommendations of particularly erotic authors. (Unlike me, she seemed to know what she was doing; at least she pulled them off the shelf with authority, and had a good stack of books in her arms.) She disowned any personal knowledge of which authors were the racy ones, but suggested that covers would be a good guide. Look for the partially clothed, muscular male bodies, she advised. This advice made perfect sense; if you can't show a hard penis on the cover, show a hard body. And indeed, Susan Johnson's *Brazen*, the very first book I opened when I got home, made good on the promise of the taut torso on its cover. Actually, the book went way beyond what I had expected from a "romance":

Angela touch[ed] the quivering tip arched against his belly, the violent throbbing between her legs matching the pulsebeat in the sharply conspicuous veins of his penis. Tracing his erection in a languid downward motion, she felt her body open in welcome as if her fingertips were vetting agent for her brain.

"Will I fit?" he whispered, sliding his hand between her thighs. "Do you think you can take it . . . all?" he softly inquired, slipping one finger into her pulsing cleft.

"I'll try," she breathily responded, the words half-swallowed as he slid a second finger inside . . .

"I want this," she finally answered on a small, caught breath, touching Kit's splendid arousal. And bending, she licked the full-stretched head with her tongue.

His fingers slid from her body. "It's yours," he whispered, watching her gently stroke his glistening length as if gauging how much she could swallow, bringing added dimension to his arousal.

"Oh, my," she uttered in fascinated admiration as the rigid shaft grew under her hands.

This passage was not exceptional. *Brazen* is loaded with "rampant" and "rigid" erections, "engorged," "stiff," and "pulsing" penises with "broad, swollen heads," and exhibits an emphasis on size ("She'd never seen such a formidable length and thickness or felt one as stiff") that belies the old, male-reassuring homilies. All of this raw sex is "redeemed," of course, by the romantic context in which it occurs. The throbbing member is tethered in the romance novel to adoration of the heroine; these men fall utterly under the emotional power of the women they worship. This is what distinguishes the romance novel from pornography for its devoted readers: the man is in love, desperately in love.

The men's movement too loves manhood-intoxicated imagery, but unlike the romance novel, encourages straight men to reclaim it for themselves—not for women. ("To become a man," says movement guru Sam Keen in *A Fire in the Belly*, "we must first leave woman behind.") Looking in the mirror after a night of sex, he describes "the image of a man I had never seen before—his cock, resting but proud, pulsated with life, his chest swelled with the joy of being, his sinuous muscles were full of power." Proud. Pulsating. Swollen. Full of power. Keen's cock itself may be "resting," but don't be fooled—his consciousness is clearly still in a burstingly tumescent state.

As these metaphors reveal, a lot of our ideas about the penis clearly come not from anatomical fact but from our cultural imagination. We shouldn't go too far, however, in the direction of viewing the "proud," erect member as entirely a human invention. Human beings may be the only creatures to put it in writing, but we are not the only creatures to wax metaphorical about the potency of the male body. Many primates augment themselves as a means of enhancing their sexual attractiveness or advertising their dominance—they fan out, blow up, stand their fur on end, expand their bodies or body parts to appear larger than they normally are. The penises of high-ranking baboons are more turgid, and thus hang farther out, than those of their subordinates.

Male chimps display their erect penises to females as sexual entice-
ment. Both reptile and mammalian penises exhibit an amazing range
of decorative enhancements to make penises appear more substantial,
flourishes that George Hersey has aptly described as "biofantasies."

The "proud member" and "throbbing manhood"—and those new
underwear ads too—are arguably among *our* species' biofantasies.
Like the decorative flourishes and augmentations of the penises of
other animals, they enhance the body part with a promise (in this case
with stimulating words) of sexual satisfaction or power. But, like
everything else that is human, our biometaphors are a product of both
nature and culture, and transcend purely reproductive function. Proud
members are as much—actually, more—a fetish of male homosexual
as heterosexual imagery, and have a place (although a controversial
one) within lesbian culture too in dildoes that are designed to look like
actual penises. Our cultural symbols do not become deactivated simply
because they are not serving some reproductive function. And, as is the
case with other primates—for whom penile display can serve not only
in sexual courtship but to establish dominance—the same bodily gestures and postures may function in a variety of ways for us too. Human men use their proud members to intimidate competitors, not just
to attract sexual partners, as we'll later see.

While celebrated in gay male pornography and men's movement literature, the "throbbing member" seems to have little appeal for most
young men. I was fascinated to read, in a study of the terms that
American college students give for the penis, that while the "throbbing
manhood" genre of penis metaphor was prominent among the group
of female students questioned, it had no equivalent in the male group.
The male group, on the other hand, had several popular categories
that were absent from the women's conceptual scheme: names
of heroic or mythic status ("Jupiter," "Genghis Khan," "The Lone
Ranger"), tools (garden hose, crank, gearshift, jackhammer, and so
on), and weaponry, both dangerous (torpedo, stealth bomber, pistol)
and innocuous (squirt gun).

This difference makes sense to me, and not because I think that
young men are inherently violent. In romance novels like *Brazen*, the
throbbing member appears in the service of giving pleasure to those
who yearn for it or need to be awakened by it—a scary assignment for
most young men. What if one can't rise to the occasion? No wonder

they prefer mechanical penile metaphors. Big rig. Blowtorch. Bolt. Cockpit. Crank. Crowbar. Destroyer. Dipstick. Drill. Engine. Hammer. Hand tool. Hardware. Hose. Power tool. Rod. Torpedo. Rocket. Spear. Such slang is violent, yes, in what it suggests the erect penis can do to another, "softer" body. But perhaps this aggressiveness is a pre-emptive strike, not against women's bodies but against becoming a "soft" body oneself.

I realize the idea of a metaphor offering protection is strange. The protection comes at the level of imaginative, not material, reality. Think about how the machine metaphors encase the penis in various sorts of metal or steel armor, making it a kind of cyborg and suggest-ing that when the penis is without such armor—that is, when "soft"— it is naked, exposed, without protection. The erect penis, these metaphors imply, has a soft being living inside itself too, like a snail within a shell. Stripped of its shell, no longer an armored warrior, it becomes a creature that can be stepped on, or (if that's going too far for you) exposed, defenseless "flesh." A human organ of flesh and blood is subject to anxiety, ambivalence, uncertainty. A torpedo, rocket, or power tool, "Jupiter" or "Genghis Khan," in contrast, would never let one down. Boys talking to each other, using these terms, can identify with a state of bravado and toughness that they don't really feel.

Interestingly, such names are often given to dildoes too. It is with some insight, then (as well as humor), that lesbian theorist Pat Califia suggests that "dildo envy" inhabits the male unconscious more than penis envy torments women. "Cocks seem more fragile than ther-monuclear to me," she writes. "There's a vulnerability about getting an erection that I'm really grateful I don't have to experience before I can give someone a night to remember." Thermonuclear nicknames like "torpedo," turning the penis into a sort of imaginary dildo (rather than a feeling, "throbbing" member), defend against that vulnerability. Dildos "don't falter or become feeble," Califia points out. "They stay up as long as the girl around them is in a mood to keep on coming." Much safer to have a torpedo as one's love tool than an organ of flesh and blood.

Armored warriors: from these fascist "comrades" to contemporary advertisements, the proud, hard body is a metaphor for mastery and power

soft races

No one wants to be a squishy snail. A classic device for convincing oneself that one is *not* something is to imagine that someone else holds the franchise on that particular quality. Throughout history, certain racial groups have been made to play the role of the squishy snail, while others can imagine that they—in contrast—are nature's select band of armored warriors. In *Male Fantasies*, Klaus Theweleit argues that for the German fascists, there existed two types of bodies: the up-standing, steel-hard, organized "machine" body of the German master and the flaccid, soft, fluid body of the Other. Here we have a picture of two phallic German "comrades" (as the statue is called). Their actual penises are small; in a sense their entire bodies play the role of proud, erect phalluses.

The Jewish man, in contrast, is represented in Nazi literature as dwarfish, womanish, simpering, impotent. In cartoons and caricatures, his body is bent, wilted, he grovels. In racist tracts, a great deal is made of the difference of his "foreshortened" penis; it's a kind of mirror image to the mythology of the black man's "Long Dong Silver." Today, blacks and Jews are seen by most people as belonging to different races. But in fact, throughout much of history, Jews and blacks were continually identified with each other—and contrasted to other races—by virtue of shared origins, shared blood, and similar appearance: full lips, large noses, dark curly hair, dark skin, and unusual genitals. Both the soft, foreshortened Jewish man and the animalistic, overendowed black man are products of the same ugly tendency of racism: to imagine the racially despised body as sexually abnormal (grosser, as with blacks, or shriveled and impotent, as with Jews) in contrast to one's own, "superior" race.

Freud even suggested that castration anxiety was "the deepest root of anti-Semitism": "In the nursery," he wrote, "little boys hear that a Jew has something cut off his penis—a piece of his penis, they think—and this gives them a right to despise Jews." Men despise women for the same reason, he argued.

"That the penis could be missing strikes [the little boy] as an uncanny and intolerable idea, and so in an attempt at a compromise he comes to the conclusion that little girls have a penis as well, only it is still very small; it will grow later. If it seems from later observations that this expectation is not realized, he has another remedy at his disposal: little girls too had a penis, but it was cut off and in its place was left a wound. . . . [T]he boy in the meantime has heard the threat that the organ which is so dear to him will be taken away from him if he shows his interest in it too plainly. Under the influence of this threat of castration he now sees the notion he has gained of the female genitals in a new light; henceforth he will tremble for his masculinity, but at the same time he will despise the unhappy creatures on whom the cruel punishment has, as he supposes, already fallen."

By Freud's logic, then, the fact that Jews are seen as belonging to an inferior, "feminine race . . . lacking in virility" stems from the notion that both Jews and women seem to have been castrated. Whether or not you believe that little boys undergo the revelations and anxieties that Freud describes (the theories have certainly received their share of criticism), it seems more likely that imagining that circumcision is a kind of castration is an effect rather than a cause of anti-Semitism. The Jew is seen as inferior, impotent, a feminine weakling to begin with—

and then those ideas are confirmed for the anti-Semite by the Jew's "mutilated" penis (circumcision is often described this way, not just in Nazi literature but in nineteenth-century sexology). See how odd, how barbaric they are? They mutilate their sons' penises! What greater proof could you have that this is an undeveloped, unenlightened race?

Freud is certainly right, however, about the cultural feminization of the Jewish man, which goes way beyond formally anti-Semitic cartoons and tracts, to include such familiar pop culture types as "passive fathers married to domineering, often vulgar Jewish women, nebbishy husbands and boyfriends involved with beautiful Gentile women; neurotic Jewish American princes; self-deprecating nerds." Jewish men are not alone, however, in constantly being called on to play the role of the ineffectual nerd; similar associations and images have dogged Asian men, perhaps even more consistently. Cultural images of Jewish men have included aggressive little dynamos like Duddy Kravitz and Sammy Gluck—characters that may evoke stereotypes of the pushy Jew but still have sexual potency. More recently, Michael Steadman (Ken Olin) of *Thirtysomething* and Jerry Seinfeld have managed to endow the neurotic Jewish prince with a certain amount of sex appeal. In contrast, Asian-American men, as actor Marc Hayashi has put it, "are the eunuchs of America." Obsequious houseboys. Mousy little scientists with Coke-bottle glasses. Even the samurai warriors and karate masters, cultural theorist Richard Fung notes (in a piece aptly titled "Looking for My Penis"), are "characterized by a desexualized Zen asceticism."

I wondered, reading Fung's pre-Jackie Chan piece, whether this was still true. I rented a few Chan movies and was startled to realize how affably asexual Chan's presence is, for all his physical stunts and feats. "Mr. Nice Guy" is what his latest hit is called. In one sense, it's meant ironically, since he's capable of beating the shit out of everyone else in the film. But in another sense, he really *is* Mr. Nice Guy: polite, sweet, and oh, so respectful of women. Even when he's fighting, he never grimaces or seethes like other action heroes. When Asian men are *not* desexualized in this way, their desire is made to seem pathetic, deluded, ridiculous. Who can watch *Breakfast at Tiffany's* today without flinching at Mickey Rooney as Holly Golightly's ("Miss Gorrightry's") lecherous, buck-toothed neighbor? Although physically less of a caricature, the Japanese-American man in *Fargo*, as David Mura argues in an Opinion piece in *The New York Times*, is in the same line of ineffectual characters, leering at white women but completely without sexual authority or magnetism. (The *Fargo* character lies about his wife's cancer in a clumsy, inappropriate pass at extremely pregnant Marge, who's an ex-schoolmate of his.) No wonder actor Chow Yun-Fat's sex appeal was noted with amazement by Western critics when he made his American debut in *The Replacement Killers*. The (rather Western-looking) Fat was arguably the first Asian actor permitted to *be* sexy, to *be* "manly," in a Hollywood movie.

Ethnic performers and writers have sometimes contributed to the stereotypes that dog them. We've seen this in the self-deprecating sitcoms and stand-up routines developed by Jewish comics, the Sambo-like characters of some black comedians, the violence of gangsta rap. Partly, these routines and characters play to dominant expectations; they make people laugh or stir their emotions, and entertainers want that. Working with—rather than against—stereotypes may also be a way to get the jump on racism and anti-Semitism, like fat people who tell the fat jokes before others tell them about them. But, too, there is often that pungent, incisive bit of *cultural* and historical (as opposed to racial) truth in the stereotypes, which groups recognize, identify with, and cannot tell their own story without.

Unfortunately, although whatever validity the caricatures have is partial, they are gobbled up by consumers as the whole, essential truth of a people. People howled (or cringed) at Philip Roth's portrait of domineering Jewish mother Sophie Portnoy who stood over her son with a knife to make him eat. They were less apt to remember Alex's more loving, lyrical (and, admittedly, less frequent) memories of his father and mother. But rereading *Portnoy's Complaint* today, thinking about the male body, I am less struck by the portrait of Sophie Portnoy than I am by Alex/Roth's ambivalence about his kindly, anxious, powerless Jewish father, whose forever constipated bowels are "doomed to be obstructed by this Holy Protestant Empire" but whose manly member has miraculously escaped cultural castration:

"Pregnable (putting it mildly) as his masculinity was in the world of *goyim* with golden hair and silver tongues, between his legs (God bless my father!) he was constructed like a man of consequence, two big healthy balls such as a king would be proud to put on display, and a *shlong* of magisterial length and girth. And they were *his*: his, of this I am absolutely certain, they hung down off of, they were connected on to, they could not be taken away from, *him*!"

Portnoy almost immediately goes on to complain about how his mother nearly took away his own balls (psychologically speaking) by sending him to the store for Kotex, holding a knife over him when he wouldn't eat, and so on. In *Portnoy's Complaint*, women, especially Sophie Portnoy, take the lion's share of Alex's kvetching. Even the "clinical" definition of "Portnoy's Complaint," provided by Alex's doctor O. Spielvogel, in a joke definition at the front of the book, is linked to castration fears which are "traced to the bonds obtaining in the mother-child relationship." By making Sophie the Chief Emasculator of the story, Roth reduces the moments in which Alex starts to tell another story—about the disempowerment of his father, not by his wife, but by the Gentile masters of the universe—to mere grace notes. Grace notes, but definitely there. Alex writes about Jack Portnoy: "The self-confidence and the cunning, the imperiousness and the contacts, all that enabled the blond and blue-eyed of his generation to lead, to inspire, to command, if need be to oppress—he could not summon a hundredth part of it. How could he oppress?—he *was* the oppressed. How could he wield power?—he *was* the powerless."

A moment later, he's quickly on to complaints about the "gender imbalance" in the household—"if my father had only been my mother! and my mother my father! But what a mix-up of the sexes in our house!" It's Sophie's unnatural dominance rather than cultural marginalization, after all, that has unmanned his father. Yet that masterful *shlong* remains, described in language that seems almost a direct answer to the typical anti-Semitic depictions: "*Shlong*: The word somehow catches exactly the brutishness, the *meatishness*, that I admire so, the sheer mindless, weighty, and unself-conscious dangle of that living piece of hose through which he passes streams of water as thick and strong as rope." Every adjective in this description has a mirror-image counterpart in anti-Semitic stereotypes about Jews as overintellectual, weak, morbidly, obsessively hyperconscious, and, of course, effeminate. Portnoy's descriptions of his father's penis offer a poignant, true—and unfortunately muted—counterdefinition of "Portnoy's Complaint" to the one offered at the front of the book. It's not Sophie and her Jewish mama's knife that are the real monsters in the story of the Jewish man's struggle to achieve manliness, not Sophie who caused the wound. She's a victim of that wound herself.

Roth implicitly acknowledges this in the kind of female characters his protagonists are attracted to. In his earlier books, they're sexy, athletic, non Sophie-like Jewish girls like Brenda Patimkin ("sweating princesses," as Riv-Ellen Prell calls them), later they are actual Gentiles. All lavish uninhibited adoration on his alter egos' penises. In *My Life as a Man*, Philip Roth creates a teenage girlfriend for Nathaniel Zuckerman. She's Sharon Shatzky, a wealthy, "rangy Amazon" of a Jewish princess, who writes him letters full of adoration for his member:

> Dearest dearest all I could think about while playing tennis in gym class was getting down on my hands and knees and crawling across the room toward your prick and then pressing your prick against my face I love it with your prick in my face just pressing your prick against my cheeks my lips my tongue my nose my eyes my ears wrapping your gorgeous prick in my hair.

Philip Roth and I attended the same Newark, New Jersey, high school. I knew the Sharon Shatzkys and Brenda Patimkins of the pastoral suburbs outside the grimy city. Perhaps Roth—although he attended Weequahic High after World War II, in an era less experimental than the sixties, in which I came of age—had discovered a hidden cache of long-limbed, dirty-talking Jewish girls whose sexual attitudes were astonishingly different from those of the rich girls I knew. Or perhaps his female characters, to borrow a phrase that Roth himself uses to title the story in which Sharon Shatzky appears, are "Useful Fictions" for the author as well as his male protagonists. Because Roth invests women with the awesome power to wound Jewish manliness, they become the only ones who can heal him—his women are not wounded (read: not Jewish) themselves, and they miraculously adore *him* (that is, his fictional alter-egos) all the same. (As David Mura describes the same dynamics in his own youthful obsession with non-Asian women: "I thought if I was with a white woman, then I would be as 'good' as a white guy." Gay men sometimes feel the same way about "straight-looking" lovers.)

As a Jewish woman, I resented Roth's long-limbed fantasy creatures. But I recognized the syndrome all too well in my own erotic preferences. When I was in college, all my dates looked like Ashley Wilkes from *Gone With the Wind*.

the cult of hardness

To be exposed as "soft" at the core is one of the worst things a man can suffer in this culture. Tears are permissible, even admirable, when they fill the eyes of an old warrior reminiscing about battle or a jock talking about his teammates. In such contexts, tears are like the soft penis after satisfying sex: they don't demean the man but make him lovable and human—because he has proved his strong, manly core. If a man is seen as soft at the core—as, for example, Bill Clinton has been—he is permitted much less latitude, and constantly has to prove that he can "play hardball," "take a firm stand," and so on. During the campaign and his first year in office, the press would seize mercilessly on Clinton's doughy physique, as though his soft, undisciplined body and taste for French fries—immortalized in a *Saturday Night Live* skit which had him jogging from one fast-food stop to another—exposed just how "unpresidential" he was. Some people seemed to appreciate Clinton's lack of rigidity and control over his body. Camille Paglia wrote in the on-line magazine *Salon* that "with his hamburgers, horse laughs, flirtations, schmoozing, and easy tears," Clinton symbolized "emotional openness and enjoyment of life." But for many, Clinton's body showed he was not a "real man"—like the mean, lean, skydiving George Bush or the horseback-riding Ronald Reagan. Those guys didn't *have* to run every day, like some wimpy yuppie; their regular "manly" activities kept them hard.

Clinton's body is a lot leaner now, and he's proven he can "play hardball" on the international scene and in domestic politics. That doesn't prevent journalists from processing whatever trouble he gets himself into through emasculating, infantilizing metaphors and tropes of voracious appetite. Joe Klein, commenting in *The New Yorker* on Clinton's behavior in the Lewinsky affair, describes Clinton's "endless chatter" with friends as always having had "an odd, distended, needy quality." (I believe this is the first time I've seen the word "distended" applied to a conversation.) He laments that the public never got to see how "adolescent" Bill really is: "the temper tantrums, the almost hilarious self-involvement, the solipsistic sense of entitlement which led him to indulge his appetites with a White House intern." John Kennedy, it's well known by now, was far more rapacious and reckless than Clinton when it came to matters of sexual entitlement. Yet even in 1998, I've not yet heard his behavior described as "adolescent" or an "indulgence of appetite." I guess it's more dignified to sneak around with Mafia molls and pricey prostitutes than interns. Or perhaps Kennedy's Brahmin bearing and cool, ironic presence—can you imagine *his* eyes welling with tears?—combined with his well-advertised wartime heroism will forever protect him from ridicule. No matter how many Kennedy scandals we unearth, in the cultural imagination he will never be the undisciplined, needy, hungry little boy that Bill Clinton is.

Real little boys, as psychologist William Pollack writes, will do almost anything they can to avoid being seen as soft. He recalls a friend's son, participating in one of his first Little League games, being hit in the head by a hardball. Looking like a "miniature punch-drunk fighter about to go down for the count," he mustered the composure to tell his mother, who had run out on the field to comfort him, to let him be. "Not here, Mom," he had whispered to her, fighting back tears. "Big guys don't cry on the field."

Sometimes, parents will encourage their sons to "toughen" up. Pollack recounts the story of how Olympic gold medal winner Oscar De La Hoya became a boxer:

"During his third birthday party, Oscar became frightened by the violence of the traditional piñata game. This is the game in which he and each of his friends were blindfolded and then, using a long wooden cane, were asked to take turns whacking the multi-colored toy-stuffed doll that was suspended above them by a cord. 'I got scared,' Oscar remembered. 'I started to cry hysterically and ran away in panic.' Oscar's parents threatened him and then punished him, but nothing could get him back to that fearsome scene. Later his father saw Oscar fleeing from other boys when they threatened to punch him. His father felt that Oscar's lack of manliness was a 'disgrace,' a shame upon the family. The 'best medicine' for his son, he felt, was to teach him to box.

After all, that's what Oscar's grandfather had done one generation earlier with Oscar's father, when he too had seemed 'unmanly.' "

The first time La Hoya boxed, he was punched badly smack on the nose, and ran home in tears. But, as he reports, he soon "learned to manage" his fears. Since he went on to become a prize-winning boxer, this initiation could be read as the prelude to a success story. But a price is paid for the "hardening" of boys (as Pollack calls it): they learn to become anesthetized to both physical and emotional pain and to keep it to themselves. He cites studies which show that by the time a boy reaches junior high school, one in ten of them has been kicked in the groin—yet the majority never tell an adult about it. Girls are ashamed to tell others when they are raped or abused, often secretly feeling it to be their own fault. Boys are ashamed to tell others when they are injured; the simple act of telling is an admission that they are not bearing their pain silently, stoically, "like a man."

Both boys and girls, when abused or shamed, often turn to their bodies in an attempt to establish a private domain in which a sense of control and self-esteem can be reestablished. Girls may go on a strict diet and exercise regime, too often escalating into a serious eating disorder. Boys, more typically, will turn to bodybuilding (something more girls are doing nowadays too). For neither girls nor boys is this just about "looking good." It's about developing a body that makes one feel safe, respected, in control.

Our aesthetic ideals, no less than our sexual responses, are never just "physical." In our culture, the hard body is a "take no shit" body. Have you ever seen an advertisement that displays a muscled torso *and* a smiling, warm face? The broad grinning faces of the competitors in bodybuilding shows seem misplaced, stuck on the wrong physique. On television, that "BowFlex" guy just won't crack a smile, even standing in front of the silliest-looking exercise machine I've ever seen: "Yes, our equipment may look a little different. That's because we've designed it to *function correctly*." Ads for women's exercise equipment capitalize on the equation of muscles and toughness. "A man who wants something soft and cuddly to hold should buy a teddy bear," declares a Reebok ad, its kick-boxing, grimacing model looking like she will take no prisoners. For women like me, definitely on the soft end of the spectrum, our bodies are not only out-of-date but politically incorrect: regressive, unacceptably "feminine."

If we do transform ourselves, the culture rewards us for our new "hardbodies," and not only with dates but also with respect. Once, one summer, after a long-standing and important relationship ended, I found myself drawn into what began as a bit of casual weight training but which within a few weeks became a daily compulsion, lasting for longer and longer periods each day. I remember very distinctly the thrill of flexing the thighs which had shamed me with their looseness, and seeing muscles tighten in them. I also remember the way having definition in my upper body made me feel when I walked past men— not just more attractive, but powerful. Their eyes did not penetrate me and reduce me to something smaller and weaker, but glanced over me with admiration, as though I were an equal. The feeling, I am certain, had nothing (or little) to do with a real change in how men saw me. My pecs were never *that* developed; by today's standards they were marginally defined. Rather, my feeling of invulnerability and power had everything to do with my having banished my then-hurting femininity from my body. I no longer felt that my body revealed my soft, bruised feelings, but instead radiated independence, toughness, emotional imperviousness.

Gay theorist Ron Long describes a similar motivation behind the muscle craze among gay men. It's not just about looking good, but about dispelling homosexual stereotypes, by embodying an ideal of masculinity which announces that one is a real man whether or not one is a "top" or a "bottom." The "butch bottom," as he calls it, "does not stake its claim to manhood on penetrating another person's body" but on being a certain *kind* of body itself. An inviolable body, whether or not it's in a "masculine" or "feminine" sexual posture. A body that challenges the cultural gaze that has cast the gay man as soft and effeminate by presenting a surface that nothing can penetrate, granite chiseled according to its owner's specifications. Similarly, when Jewish wrestler Bill Goldberg jackhammered Hulk Hogan to take the World Championship Wrestling title, "it opened a new chapter in Jewish-American history," according to Rabbi Irwin Kula. "What this says," the rabbi explains, is: "Look at us! We're not ververbalized! We're not weak or wimpy! We're the heavyweight champ!"

I resist getting hard nowadays, and not only because I lack (although I *do* lack) the sustained discipline that my gym-going friends have. Walking past men that summer, I felt proud, beautiful . . . and armored. That was the original point of my weight-lifting regimen, after all, to protect my hurting self from further wounds, wrapping it in a body that was tougher, tighter, more impervious to assault. Nothing wrong with that. I needed some armor. But there is also a time to take the armor off. A culture that idealizes, fetishizes, is addicted to the hard and impenetrable, is a cold and unforgiving place to be. In my own way, I resist becoming "one" with that culture. At the same time, I pay a price. I walk down the street with my hardbody girlfriends, and beside them feel dumpy, lowly, a snail without a shell—easily ignored and easily squished. At such moments I feel that perhaps after all we are right to keep ourselves protected. We have given men (and, increasingly, women) only two choices in this culture: soft and hard, snail or shell. It seems an awfully limited repertoire for a person, even for a bodily organ.

pills and power tools

Impotence. The word rings with disgrace and humiliation. (Philip Lopate, in an essay on his body, writes that merely to say the word out loud makes him nervous.) Unlike other disorders, impotence implicates the whole man, not merely the body part. *He is impotent.* Would we ever say about a person with a headache, *"He is a headache"*? Yet this is just what we do with impotence, as Warren Farrell notes. "We make no attempt to separate impotence from the total personality," writes Farrell. "Then, we expect the personality to perform like a machine." "Potency" means power. So I guess it's correct to say that the machine we expect men to perform like is a power tool.

Contemporary urologists have taken the metaphor of man the machine even further. Erectile functioning is "all hydraulics," says Irwin Goldstein of the Boston University Medical Center, scorning a previous generation of researchers who stressed psychological issues. Goldstein was quoted in a November 1997 *Newsweek* cover story called "The New Science of IMPOTENCE," announcing the dawn of the age of Viagra. At the time, the trade name meant little to the casual reader. What caught my eye were the contradictory messages. On the one hand that ugly shame-inducing word "IMPOTENCE" was emblazoned throughout the piece. On the other hand, we were told in

equally bold letters that science was "REBUILDING THE MALE MA-
CHINE." If it's all a matter of fluid dynamics, I thought, why keep
the term "impotent," whose definitions (according to Webster's
Unabridged) are: "want of power," "weakness," "lack of effectiveness,
helplessness," and (only lastly) "lack of ability to engage in sexual intercourse"? In keeping the term "impotence," I figured, the drug companies would get to have it both ways: reduce a complex human
condition to a matter of chemistry, while keeping the old shame machine working, helping to assure the flow of men to their doors.

It's remarkable, really, when you think about it, that "impotence"
remained a common nomenclature among medical researchers (instead
of the more forgiving, if medicalized, "erectile dysfunction") for so
long. *Frigidity*—with its suggestion that the woman is "cold," like
some barren tundra—went by the board a long while ago. But "impotence," no less loaded with ugly gender implications, remained the
term of choice—not only for journalists but also for doctors—throughout all of the early reportage on Viagra. "THE POTENCY PILL,"
Time magazine called it, in its May 4, 1998, issue, three weeks after
Viagra went on sale, breaking all records for "fastest take-off" of a
new drug that the Rite-Aid chain had seen. At the same time, inside the
magazine, fancy charts with colored arrows, zigzags, triangles, circles,
and boxes show us "How Viagra Works," a cartoonlike hot dog the
only suggestion that a penis is involved in any of this.

The drug companies eventually realized that "impotence" was as
politically incorrect as "frigidity." They also, apparently, began to
worry about the reputation that Viagra was getting as a magic bullet
that could produce rampant erections out of thin air. Pfizer's current
ad for Viagra announces "A pill that helps men with erectile dysfunction respond again." *Respond.* The word attempts to create a counter-
image not only to the early magic-bullet hype but also to the curious
absence of partners in men's descriptions of the effects of the drug. *It's*
"Stronger." *It's* "Harder." "Longer-lasting." "Better quality." *It's*
"Firmer." The characters in the drama of Viagra were three: a man, his
blessed power pill, and his restored power tool.

The way Viagra is supposed to work—as the Pfizer ad goes on to
say—is by helping you to "achieve erections the natural way—in response to sexual stimulation." *Natural. Response.* It illustrates its
themes with a middle-aged man in a suit dipping his gray-haired part-

ner, smiling ecstatically. *Partners. Happy partners.* A playful, joyous moment. "Let the dance begin," announces Pfizer at the very bottom of the ad, as though it were orchestrating a timeless, ritual coupling. The way men *had* been talking about the effects of Viagra, that dance was entirely between them and their members.

It wasn't a playful rumba, though. More like a march performed to the finale of the 1812 Overture, accompanied with cannon blasts. "This little pill is like a package of dynamite," says one user. "Turned into a monster," says another, with pleasure. "You just keep going all night. The performance is unbelievable," said one. I'm not making fun of these responses; I find them depressing. The men's explosive pride, to me, is indicative of how small and snail-like these men had felt before, and the extravagant relief now felt at becoming a "real man," imagined in these comments as some kind of monster Energizer Bunny (pardon me, *Rabbit*).

Something else is revealed too by the absence of partners in these descriptions of the effects of Viagra. The first "sex life" of most men in our culture—and a powerful relationship that often continues throughout their lives—involves a male, his member, and a magazine (or some other set of images seemingly designed with the male libido in mind). Given the fast-trigger nature of adolescent sexuality, it doesn't take much; indeed, it sometimes seems to the teenage boy as though everything female has been put on the face of the earth just to get men hot. Philip Roth's descriptions of Alex Portnoy masturbating at the sight of his sister's bra, capable of getting a hard-on even at the sound of the *word* "panties," are hilarious—and true to life. Despite myths to the contrary, it's sometimes not so different for adolescent girls either (remember my science teacher with that hot wallet in his pants). We tend to be less compulsive masturbators, though. More significantly, when we grow up, those "hard-on" moments (if, to make a point, I may use that metaphor in a unisexual way) aren't transformed into launch-off preparations for a sexual "performance."

It's often been noted that women's sexual readiness can be subtle to read. We are not required to cross a dramatic dividing line in order to engage in intercourse. And we aren't expected—as men are expected, as men seem to expect of themselves—to retain that hair-trigger sexuality of adolescence. Quite the opposite, in fact; the mythology about women is that we're "slow cookers" when it comes to sex. Men, in contrast, get hit with a double whammy: they feel that they have to perform and they expect themselves to do so at the mere sight of a fancy brassiere! It's one reason, I think, why so many men "trade up" for younger partners as they get older; they're looking for that quick sexual fix of adolescence. It's their paradigm of sexual response, their criterion (ironically, since it represents the behavior of a fifteen-year-old) of manliness.

It comes as no surprise, then, to learn that, as sexual "performers," many men seem to expect no tactile help from the audience except—hopefully—applause at the end. Gail Sheehy (who, by the way, has cleaned up her own terminology; in an article from the early nineties, her term for "male menopause" was "viropause," now she calls it "manopause") reports that Lenore Tiefer's interviews with hundreds of cops, firemen, sanitation workers, and blue-collar workers at Montefiore Medical Center in the Bronx revealed that most of these men expect, even in their fifties, to be able to get an erection just from paging through *Playboy*. Sheehy goes on: "When the sexologist suggests that at this age a man often needs physical stimulation they balk: 'C'mon, Doc, it's not *masculine* for a woman to have to get it up for me.' Their wives often echo that rigid code: '*He* should get it up.' "

Some dance, huh?

Most studies of Viagra's "effectiveness" leave partners out of the picture too. When you put them in, you get a somewhat different picture of the "success" of the drug. In England, they used something called a "RigiScan" to measure the penis's "resistance" against a cloth-covered ring while Viagra-treated men watched porn movies. In the United States, the 69 percent "success rate" that Pfizer submitted to the FDA was based on questionnaires filled out by patients. "Real soft data, no pun intended," William Steers, Chief of Urology at the University of Virginia, was quoted as saying in the July 6, 1998, *New Yorker*. He went on to note that Pfizer's study included no spousal questionnaires. Steers, cheers to him, *did* ask spouses. It turns out that when you ask women about sex with their "Viagra-enhanced" husbands, their estimation of the success of the drug is always lower than men's—about 48 percent for the women, no matter what measure of "success" you use.

Steers does not provide detail as to what those different measures were. Perhaps "monsters" were not what partners were looking for in bed. Perhaps they didn't appreciate the next-day chafing that usually accompanies "going all night." (I once knew a man who could stay erect all night; it had its upside—so to speak—but I couldn't escape the feeling that a bit of sadism, at the very least a control complex, was fueling his unbending passion.) Perhaps partners didn't want *just* a proud member, but the kind of romantic attention that goes along with the proud member in the romance novels. "We're a very meat-and-potatoes culture," says Karen Martin, a sex therapist in upstate New York. "In other cultures, they toss in a few mushrooms." Maybe the wives of Viagrans wanted a few mushrooms tossed in with the beef.

They are less apt to be served them, however, if the couple has had sexual problems. Such partners may have grown distant from each other, may no longer know how to communicate intimately with each other, physically or verbally. Since the initial wave of enthusiasm about Viagra, therapists have begun to worry that Viagra is providing couples with a way to sidestep dealing not only with relationship problems that may have *contributed* to their sexual difficulties ("just because there is a physiological problem doesn't mean there is no psychological cause," reminds Eileen Palace, director of the Center for Sexual Health at Tulane) but also with patterns of alienation, resentment, and anger that may develop *because of* those difficulties. A number of studies have found that when men begin to have erectile difficulties, a common response is to turn away from *all* romantic and affectionate gestures—kissing, caressing, hugging—so as not to (as one said) "stir things up." Many don't offer manual or oral stimulation in place of intercourse, because that would be to admit to themselves that they can't "perform" the "way a man should." They're often uncomfortable talking about the situation with their wives (and even their doctors, who report that most of the men who are asking for prescriptions for Viagra never mentioned their dysfunction before). "I would tend to kind of brush the problem under the rug," says one man. "It isn't an easy topic to deal with. It goes to the heart of your masculinity."

Into the middle of all this distance, confusion, anxiety, and strain walks Viagra, and with it the news that the problem is only a malfunctioning hydraulic system—which, like any broken machinery, can be fixed. And "let the dance begin!" Many couples, unsurprisingly, don't know the steps, stumble and step all over each other's toes.

Let me make it clear that I have no desire to withhold Viagra from the many men who have been deprived of the ability to get an erection by accidents, diabetes, cancer, and other misfortunes to which the flesh—or psyche—is heir. I would, however, like CNN and *Time to* spend a fraction of the time they devote to describing "how Viagra cures" to thinking about that gentleman's astute comment—"It goes to the heart of your masculinity"—and perhaps devoting a few features to exploring the functioning of *that* body part as well. The "heart of masculinity" isn't a mechanical pump, and in imagining the penis as such, Viagran science actually administers more of the poison it claims to counteract.

rethinking metaphors for manhood

Most of our metaphors for the penis, as you will recall, actually turn it into some species of dildo: stiff torpedoes, wands, and rods that never get soft, always perform. These metaphors, I suggested, may be a defense against fears of being too soft, physically and emotionally. But at the same time as these metaphors "defend" men as they joke with each other in bars or—more hatefully—act as a misogynist salve for past or imaginary humiliations, they also set men up for failure. For men don't really have torpedoes or rods or heroic avengers between their legs. They have penises. And penises, like the rest of the human body and unlike dildoes, *feel* things. The only one of our cultural metaphors that seems to acknowledge this fact is the "throbbing manhood." The throbbing manhood, however, is only permitted to feel *one* thing: throbbing, unabated passion. As such, he too is really a phallic dildo in disguise.

Thinking about man as a "machine," as Viagran science now does, is really not much different, is it? The penis is *not* a machine. It's not always "malfunctioning" simply because it doesn't want to—or isn't able to—perform like a power tool. I like Philip Lopate's epistemological metaphor for the penis much better. Over the years he has come to appreciate, he writes, that his penis has its "own specialized form of intelligence." It's not an "idiot brother"—it knows when it's been insulted or pushed around, when a request for sex is really a request for something else (proof of love, proof of manhood, proof to one's partner that she's still desirable, that she's not too fat to love) or when unspoken hostilities are seeking forgetfulness in the blind passion of a moment. The penis knows, too, that it is not a torpedo, no matter what a culture expects of it or what drugs are coursing through its blood vessels.

I recommend, too, that we rethink that term "hard" as it refers to the erect penis. When we imagine the erect penis as "hard" we endow it with armor. It's time to take that metaphorical armor off, not to expose a squishy snail beneath, but to begin to think of the male body in terms of its varied feelings rather than an imagined ideal of constancy. The penis, far from being an impenetrable knight in armor, in fact wears its heart on its sleeve. That's what's so magical about it. What other feature of the human body is as capable of making the upwelling of desire, the overtaking of the body by desire, so manifest to another? Perhaps, with a woman's body, "getting wet." But most other spontaneous physiological manifestations of desire are rather subtle, and easy to hide (or exaggerate, as the case may be). Many can be hard to distinguish from other states. Fear, for example, can make one short of breath. Sartre says that a certain "heavy tranquillity"—"the eyes fixed and half closed, movements stamped with a heavy and sticky, gentle sweetness"—is the unmistakable appearance of desire. But showing such symptoms I once had a man ask if I was falling asleep!

Glances, gestures, utterances, of course, can all communicate desire to another. But what we express with our eyes and hands and mouths rarely is able to signify the pure ascendancy of desire over us, for what we "say" with our eyes and hands and mouths is usually heavily mediated by cultural vocabulary, perhaps especially nowadays. We learn what sexual arousal looks and sounds like from the movies, and—as with any other language—we pick up the grammar and syntax without being aware of it. The more conventional and stylized the language, the more opaque we remain to each other. Desire transforms us profoundly "inside," alters the color, the smell, the temperature of the world for us, changes our experience of our bodies, commands us into a different mode. But the body as it appears to others in the world—the only body from which other people can draw their knowledge of us—frequently acts as mask rather than revelation of this magical transformation.

The penis has a unique ability to make erotic feeling visible and apparent to the other person, a transparency of response that can be profoundly sexually moving and empowering to the one who has stirred the response. Necking with Bobby Cohen, I felt my first bolt of sexual heat not getting "felt up" but touching *him*, and finding that he was hard. In those days, when boys touched *me* I experienced it, paradoxically but correctly, as really about *them* as sexual subjects. When I was slow dancing with Jeffrey Schwartz, I had vaguely heard, from across the room, a whisper directed at him from a bunch of boys sitting by the bar: "Did you feel it?" "It" was my breast, and actually he *hadn't* actually felt "it." (I was twelve. I wanted to look like Annette Funicello. Prosthesis was required.) That, however, was irrelevant to Jeff's victory. Certainly, whether or not *I* had felt anything (I hadn't, through all that spongy stuff) made no difference. Sherman was marching and had taken a key city. Word went quickly around school that he had "felt me up." Boys seemed much more interested in this sort of thing than in getting *us* to feel them up.

When I touched *them*, I became a subject myself, and discovered not only their sexual power but my own. "And he comes to *me*!" We can imagine Connie Chatterley's response to Mellors along the lines of a starstruck worshipper who cannot believe her good fortune at being chosen by a god. That would be the phallocentric reading of Connie's exclamation and Lawrence certainly does everything he can to encourage it. But we can also understand Connie as dazzled by her *own* power, not Mellors's—her power, not to snare the phallus (as in Eve, Medusa, Helen, etc.), but to cause the mundane instrumentality of another person's body to awaken, in her presence and because of her presence.

Sartre calls this awakening of the other person's body "the caress," emphasizing that this is a stroking that may be done with hands or eyes. To take delight in such magic powers is, of course, not specific to women (although so long as boys like Jeffrey Schwartz are intent on "scoring" body parts, they're in a whole other ballpark). Rather, according to Sartre, this is the very essence of the "dance" of desire: the mutual recognition that another's consciousness has given itself over to the body, *become* body, for oneself. When this happens, the other person's freedom—a source of perpetual anxiety, according to Sartre—is temporarily "ensnared," and so too is one's own. Desire stops all other possibilities dead in their tracks, insists that we pay attention only to the feelings aroused in our bodies. For a brief moment, our existence no longer seems *de trop*, without justification, weightless with what Kundera has called "the unbearable lightness of being." Desire claims consciousness, and endows it with the weight of the body.

Sartre's phenomenological descriptions of "the caress" are wonderful—but his examples are definitely those of a heterosexual male, and one with a particularly conventional sexuality. Casting around for illustrations of the ensnarement of consciousness by the body, he mentions tumescence in passing. But it is clearly the bringing to life of the female body that is his paradigm—caressing "the flank of a desired woman," and so forth. Penises, for him, are the things that *enter*, penetrate, they are tools, not feeling flesh. Here, Sartre's captivation by an ideology of armored masculinity has interfered with his philosophical insight (and probably his sex life, which is said to have been impressive only in quantity). Thinking through "the caress" in terms of my own experience (also biased, of course, but helpful in supplementing Sartre's), I cannot imagine a more precise and concrete illustration of the body made desire (or desire made body) than the aroused penis.

The *aroused* penis, not the "hard" penis. Armored tanks are hard. Torpedoes are hard. Members of the Master Race are hard. When people are stubborn and unbending, they are hard. That kick-boxing model, she's hard. "Here I stand," Martin Luther proclaimed in a classically phallic moment of history, "I can do no other." That was a hard stance—perhaps useful in answer to the call of history, but sex? Mellors stood there, resplendent, with his "tense phallus, that did not change." These are images that identify masculinity with unwavering, undivided constancy—with a sex which is "one." Set against that expectation, the penis and its shifting feelings may indeed appear as a rebellious "other," with a mind of its own that is less than admirable. The little brother with wet dreams. The nervous adolescent, praying he doesn't show himself to be soft, like a girl.

Arousal, by its very nature, comes and goes. It can take a variety of forms. It's not a quality of "being," but implicitly suggests relationship—someone or something that has aroused another. If we accepted this, the notion that a man requires understanding and "tolerance" when he doesn't "perform" would go by the wayside ("It's okay. It happens," still assumes that there is something to be excused). So, too, would the idea that there ought to be one model for understanding what we used to call "impotence" and now call "erectile dysfunction." Some may indeed be physiological in origin. Some may be grounded in deep psychic fears and insecurities. But sometimes, too, a man's penis may simply be instructing him that his *feelings* are not in synch with the job he's supposed to do—or with the very fact that it's a "job." Sometimes, the penis's "specialized intelligence" should be listened to rather than cured.

does size matter?

cultural perspectives on the matter of size

The young girl stands in front of the mirror. Never fat to begin with, she's been on a no-fat diet for a couple of weeks and has reached her goal weight: 115 pounds, at 5′ 4″—exactly what she should weigh, according to her doctor's chart. But goddamnit, she still looks dumpy. In her mind is this Special K commercial that she's seen a few times on television: a really pretty woman admiring herself in a slinky, short black dress, with long athletic legs, every curve perfect, lean-sexy, nothing to spare. Self-hatred and shame start to burn in the girl, and other things too. When the commercial goes on, the woman's sleek body is like a magnet for her eyes; she almost feels in love with her. But envy tears at her stomach, is enough to make her sick. She'll never look like that, no matter how much weight she loses, no matter how many hours she spends on the StairMaster. Look at that stomach of hers, see how it sticks out? Those thighs—they actually jiggle. Her butt is monstrous. She's fat, gross, a dough girl.

It's a depressingly well-documented fact that when girls and women are asked to draw their bodies or indicate their body size with their hands, they almost always overestimate how much space they take up, and tend to see themselves as too fat no matter how thin they are. This once was thought to be a "body image distortion" unique to those with anorexia nervosa. We now know that seeing oneself as "too fat" is a norm of female perception. Statistics on average weights and medical charts are irrelevant. What matters is the gap between the self and the cultural images. We measure ourselves not against an ideal of health, not even usually (although sometimes) against each other, but against created icons, fantasies made flesh. Flesh *designed* to arouse admiration, envy, desire.

I've been writing and lecturing on these female body issues for years. At almost every talk I've given, someone in the audience (mistakenly concluding that because I had talked about women, I believed they had the exclusive franchise on body insecurity) has challenged me: What about men? What about baldness? Height? Muscles? All these examples are well taken. But no one has ever brought up the more perfect analogy: men's insecurities about penis size. I myself did not realize exactly how perfect the analogy was until I read of a 1996 study in which pediatrician Peter Lee found that college men, no matter what their actual dimensions, tend to underestimate their penis size. In a mirror image of women's perception of themselves as too big (even when they are average or below average weight), men tend to see themselves as too small—even with "average"-size penises (currently defined by doctors as four inches nonerect; six inches erect).

Where do men get their ideas about how big their penises "ought" to be? Some, as the excerpts from *Portnoy's Complaint* and Tim Allen's autobiography suggest, get them from a child's-eye view of their fathers' penises. Some get them from other guys in the locker room. Some become convinced they are too small because a partner has told them they don't measure up; Scott Fitzgerald, as Hemingway recounts in A *Moveable Feast*, developed anxieties after Zelda told him (Hemingway reports how Scott described her words) "that the way I was built I could never make a woman happy . . . She said it was a matter of measurements." (Hemingway takes Scott off to the rest room to inspect him and declares him "perfectly fine," but he remains unconvinced.) But many men, like women, get their ideas about how big they should be from the bodies of cultural icons: the Dirk Digglers and Harry Reemses of video porn and sex magazines, hired specifically for their endowments. (In the late 1990s, I'm sure the guys in the underwear ads are doing their part too.)

These guys are as off-the-charts vis-à-vis average penis size as the runway model is vis-à-vis the average female body. In the August 1997 issue of *Playgirl*, with nonstop penises from cover to cover, the only man who appears to have an "average" member is Brad Pitt, who is featured in a set of paparazzi photos. I was glad to see Pitt in there—proof that a man can be incredibly sexy without being incredibly well hung. *Playgirl*, although officially edited with a female reader in mind, is sold alongside *Torso*, *Jock*, and *Hustler*; I'm sure it has a male readership among both straights and gays. In any case, I've heard (from a friend who knows some vendors) that because of Pitt, one of those men who appeal to virtually every sexual and gender orientation, this particular issue sold out as soon as it hit the stands. Unfortunately, the beauty of diversity is not exactly the message conveyed by the penis stats and descriptions listed in the magazine's "Sex International Network" want ads. "Very hard 7″ cock. Big balls." "7½″ thick penis—dark meat." "8″ penis, very long lasting. I stay hard after I come." "Love masturbating 2–3 times a day with my 8″ cock." "9½″ hard penis, very wide and ready to please any woman." "8½″, cut, rock hard cock." "11 inch cock looking for a beautiful blonde female." "7″ hard, thick cock. I stay hard all night and I know I can take care of you." And so on.

The humongous penis, like the idealized female body, is a cultural fantasy. It exists in the flesh; some men—like those featured in the photos and want ads of *Playgirl*—do have very large penises. But let's put it this way: If a Martian was planning a trip to earth and was given a *Vogue* and a *Playgirl* to enlighten him on what to expect from human women and men, he'd get a very misleading impression. So does the average male reader or viewer of porn. And even if he knows, on some level (from his experience in locker rooms and the like), that the Dirk Digglers, Harry Reemses, and Jeff Strykers of the world are not the norm, that knowledge may pale beside the power of the iconography: the meanings attached to having an impressively large member. The woman in the Special K commercial is a "real" woman too (although these images are, increasingly, digitally manipulated); it's the fact that she so perfectly, precisely embodies current notions about femininity and beauty that makes her a fantasy, and an oppressive standard for the ordinary woman to aspire to.

Think, to begin with, about that adjective "impressive," which came spontaneously to my mind as I wrote the phrase "impressively large member," and all that it conveys. We wouldn't usually describe large breasts as "impressive," would we? ("Bodacious" isn't in the dictionary and I'm not sure that I know exactly what it's supposed to mean, but it sure doesn't seem to me to suggest a body part demanding respect.) In contrast, the penis so large as to take a lover's breath away is a majestic penis, a commanding penis. From romance novels ("His strength was conspicuous beneath her hands, his muscles prominent, steel hard. He was strikingly large . . . so *very* large . . .") to the erotic fantasies in the back pages of *Playgirl* ("I watched in curiosity and amazement as he unzipped his pants, revealing a magnificent cock . . . As his manhood sprung out at me, hard and thick, I gasped and stared . . .") to Ally McBeal, the woman's first encounter with the male stud's member is typically one of gasping bedazzlement at his "magnificent" size. Gay male erotica has similar moments: "Lew's breath was stolen by Jeff's cock. Sure, the mountain man had seen a few in his time. Many. Some were as nicely shaped. A few were as tasty-looking. But none were as gigantic." Perusing this literature, I couldn't help but think (with a mental chuckle) about Freud's description of the origins of "penis envy"; little girls, he wrote, "notice the penis of a brother or playmate, strikingly visible and of large proportions." Sounds quite a bit like the rhetoric of erotica to me, with its fantasy of a penis so impressive it simply dazzles the onlooker, takes his or her breath away.

The bedazzlement need not be sexual, however. The warlords of the Ottoman Empire publicly posted their genital measurements for conquered tribes to admire. Appearing a "big man" to other men is an important aspect of men's preoccupation with size. I've personally heard three different variations on the following joke, most recently the garbled version told in the movie *Slingblade*: Three men are urinating off a bridge together. "River sure is cold," says the first man. "It's deep, too," says the second. "Sandy bottom," says the third. The joke is our contemporary version of the fresco at the Roman ruins at Pompeii (circa A.D. 79), which depicts a wealthy man using his enormous penis to counterbalance several bags of money on a scale. The big penis is worth its weight in gold, the winner in contests among men. One young man, who had his penis pierced to endow his "little dick with a lot of fucking attitude," suggests "the big-size thing develops in the school locker room when you're a kid. The big-dicked guys send out signals that say, 'We're better,' 'We're more masculine than you,' or 'We deserve to be here, look at the size of our dicks.' "

Penile augmentation is an increasingly booming business in this culture, and many of the men who have their penises enlarged do it for "display purposes." "I'd always been happy in an erect state," says one man, "I never had any complaints from my wife—but I had a lot of retraction when flaccid. It's not that I want to flaunt myself at the gym, but I didn't want to feel that self-conscious." Others do want to "flaunt"; according to surgeon Melvyn Rosenstein the typical phalloplasty patient "wants to get big so he can show himself off to other men, to say, 'Mine is bigger than yours,' like a buck deer displaying its antler." Most phalloplasty patients, doctors add, do not have especially small penises. "The overwhelming majority of men I do are unquestionably normal," says Rosenstein. "I had a guy in the office yesterday who was concerned that he was small. I assured him that he was normal, but he said, 'Let's go ahead and do it.' "

Like George Costanza in that "shrinkage" episode of *Seinfeld*, most phalloplasty patients are haunted by a humiliation that is likely only imagined. The cultural backdrop of their anxieties, however, is not imagined, any more than women's anxieties about the size of their breasts are of their own making. *"He's the nicest guy I ever dated. But he's just too small."* So reads the bold print of an ad from Dr. Gary Rheinschild, who specializes in penile augmentation and who in 1995 claimed to have performed more than 3,500 such operations. Rheinschild also uses phrases like "shower syndrome" and "locker room phobia" (to describe what the man I've quoted above suffers from) and hopes to make penis enlargement "as common as breast implants." But even before cosmetic surgeons began their campaigns, ads hawking miracle products for increasing penis size both exploited and exacerbated already existing male insecurities by drawing on the equation: penis size = manliness. "Dramatic Increase in Penis Size!" boast the makers of "NSP-270," marketed in the eighties:

"Boys who couldn't measure up to the Navy's proud standards of manhood . . . who would never be able to satisfy a 'woman in every port' . . . who would disgrace the uniform if they were ever allowed to wear it . . . were given massive dosages of this amazing sex nutrient . . . [and] suddenly and dramatically experienced: *Proud Erections! Dramatic New Ability in Intercourse! Supercharged Sperm That Now Can 'Do the Job!'* . . . And, most amazing of all, fantastic growth in penis size!"

Anxiety about "shrinkage" and size are not exclusively Western either. Southeast Asian men suffer from a form of male panic (some might call it hysteria) known as Koro, in which they imagine that their genitals are retracting into their bodies and will ultimately cause their death. Recent studies suggest that Koro is triggered by penile vasoconstriction (or "shrinkage") caused by cold, fright, and other states. Penile augmentations—making use of pins and inserts—are performed in many cultures. Groups ranging from the Caramoja tribe of northern Uganda to the sadhus of India have practiced the technique of tying weights to the penis in order to make it longer. The sadhus, who believe that God dwells in the penis, stretch themselves to lengths of twelve to eighteen inches. By contrast, John Bobbitt's boast to Jenny Jones that his reattached penis is "stronger and bigger than ever," with a fraction of an inch added on by the surgery that reattached it, seems pretty flaccid. (Bobbitt, as mentioned earlier, went on to have an augmentation which, he claims, added three inches in length, one inch in girth, and made his penis "like a beer can." Perhaps, at least in Bobbitt's case, Pat Califia is right about "dildo envy.")

Most of the transformations wrought by penile augmentation in this culture—usually, gains of a couple of inches at most—lack the clear, ritualistic drama of organs that have been augmented—like the sadhus'—into hyperbole. But symbolically, the change can be just as potent, a fact that surgeons exploit. "I get [my clients] to see this as an incredible change in their lives," Dr. Jamie Corvalaan explained to *Esquire* reporter John Taylor. "I tell them, 'This is going to change your self-image, change the way you walk, sit, look, do business, pursue women. You will now act like a man with a big penis." How does a man with a big penis act? Well, we know that he can exhibit himself with pride in the locker room. The surgeons, addressing the concerns and fantasies of men who perceive themselves as small, promote the large penis as the route to self-confidence, assertiveness, social authority. But men who are born with large penises, as I've discovered from talking to several, may experience their size as embarrassing excess rather than a cause for pride. When one of these men described his large penis to me as "a problem," my immediate reaction (thankfully, not outwardly expressed) was similar to that which I've had when very slim women complain that they just can't keep the weight on. "That's a *problem?*" But further conversation revealed that this man had indeed had problems, not only in being too large for many partners but also with an abiding sense of shame.

The fact is that human cultures have been somewhat ambivalent about very large penises. Yes, they advertise male potency, and have often functioned as symbols of reproductive fertility. But at the same time, like very large breasts, they are often viewed as gross and a sign that there is nothing much "upstairs" (the body's endowments being seen as hydraulically regulated, I guess; what accumulates at one end has been forced out the other). Long Dong Silver laughs at male proportions, turns them into so much clumsy poundage. Dong. Shlong. Alexander Portnoy aside, these are usually words of disdain, not reverence, in our culture. Also, body-part size and excessive sexuality have often been joined in our cultural imagination, and thus don't fit well with the heroic, civilized ideals that men are supposed to uphold. There are some interesting depictions of Christ with a large penis, possibly an erection. But typically, in classical Western art, the convention has been to represent the heroic body as muscular, but the actual penis as rather small. That's why my trips to the Newark Museum were so disappointing.

Ancient Greece, a highly masculinist culture but also one that placed great emphasis on male self-control in matters of sexuality, favored "small and taut" genitals. "Large sex organs," as Eva Keuls points out, "were considered coarse and ugly, and were banished to the domains of abstraction, of caricature, of satyrs, and of barbarians." In those domains, the penis was represented as often grotesquely huge, as though absorbing all the sexual excess that the "civilized" Greek would not permit to have a place in his own self-conception. White Europeans have performed the same projection onto racist stereotypes of the overendowed black superstud,[*] stereotypes, as discussed earlier, that got brought right onto the floor of the United States Senate during the Clarence Thomas/Anita Hill sexual harassment hearings. Long

---

[*]Robert Mapplethorpe's controversial photograph, "Man in a Polyester Suit," satirically illustrates this racist opposition between the white man's "civilization" and the black man's primitive endowments, by showing a gigantic organ spilling out from the unzipped fly of a black man in a tidy business suit, whose polyester material jokingly represents the tacky, Kmart artifice of "civilization."

Dong Silver is racialized pornography, of a piece with the brown dildoes that feminist writer Heather Findlay encounters in a sex-toy shop: "I turned and looked. They were not dildoes; they were *monstrosities*. Twenty-four inches and thick as my arm. 'Big Black Dick' said the wrapper . . . I looked around for some 'Big White Dick' or even 'Big Flesh Colored Dick.' No luck." Race, she concludes, "permeates American culture."

Yes, and not just American culture, but the Western psyche. Frantz Fanon, discussing the racial fantasies and dreams of his white psychiatric patients, has written: "One is no longer aware of the Negro, but only of a penis: The Negro is eclipsed. He is turned into a penis." White boys like Norman Mailer have envied this instinctual status (in *The White Negro*, he admires the Negro's "art of the primitive" and calls jazz "the music of orgasm"), and some black rappers and athletes may capitalize on it. (Jack Johnson, the first African-American world heavyweight boxing champion, wrapped his penis in gauze to emphasize its size as he paraded around the ring during his public matches.) But primitive manhood has no place under the robes of a Supreme Court Justice. There were moments when I felt deep pain for Thomas. To be on the brink of *real* respectability—perhaps the most respectable position this country can offer—and to be tailgated by Long Dong Silver! "This dirt, this sleaze," Thomas told the committee, "is destroying what it has taken me forty-three years to build." The white senators could empathize, even identify with the pain of having charges of sexual misconduct threaten to destroy one's career. But they could not bond with the racial dimension of Thomas's predicament, only look on it with horror.

Anita Hill was the first to mention Long Dong Silver. But once he had been let loose in the Senate, Thomas knew he would be dogged by him, whatever he did, so he made a bold and cunning move. Rather than allow the equation "Negro = penis" to remain unspoken, doing unconscious damage to his hopes of confirmation, Thomas drew attention to Long Dong Silver's racial overtones, thus suggesting that Thomas *himself* was the real victim—of ugly racial stereotyping. The strategy was exploitative and outrageously misplaced. Hill was a black woman, as people seemed to forget, with little to gain by resorting to racial smears; even if she had been lying, she could easily have come up with a different, equally offensive but racially neutral image. Thomas's strategy, however, proved triumphant. The senators, going out of their way to prove to the world that *they* (unlike Fanon's patients) did *not* see the black man as penis, made him a Supreme Court Justice.

yes, but does size *really* matter?

The magnificently large penis, as we've seen, is an icon of cross-cultural potency. So it's not surprising that size matters very much to men. (According to a *Glamour* magazine survey, when men were asked whether they would rather be (A) 5 feet 2 inches tall with a seven-inch penis or (B) 6 feet 2 inches tall with a three-inch penis, 63 percent of the respondents picked A and only 36 percent picked B.)

Yet there is an extensive medical, scientific, and cultural literature designed to counteract the notion that penis size "matters" to a partner's sexual pleasure, a reassurance that—although by no means scientifically rigorous or convincing—has become virtually official dogma today. I have no comparative statistics, but I suspect that because of this reactive "counterculture," far fewer contemporary men are as obsessed with, tormented by insecurities about penis size as women are about weight. "Penis size has never been shown to affect sexual pleasure for men or women"—I've read some version of this claim over and over, everywhere from popular-magazine articles to sophisticated scholarly books on the construction of gender, to medical guides. An example of the latter is Abraham Morgentaler, M.D., who says (in *The Male Body: A Physician's Guide to What Every Man Should Know About His Sexual Health*) that he offers the following story about Abraham Lincoln in an attempt to calm his patients' anxieties about size:

"President Lincoln was extremely tall for his time and was once asked his opinion on the proper length of a man's legs. His common-sense answer was that a man's legs should be long enough to reach from his hips to the ground. In similar fashion, a man's penis should be long enough to reach inside his partner."

Hmmm . . . Sounds a bit like the old sexist joke about breasts, that more than a mouthful is wasted. But many female doctors seem to agree with Morgentaler. Lenore Tiefer, associate professor of urology and psychiatry at the Albert Einstein College of Medicine, when challenged to explain why the partners of phalloplasty patients had more climaxes after the operation, offers the following: "Whatever increases a man's sexual confidence will probably make his wife happier." Without even considering it, Tiefer discounts the possibility that her husband's larger penis might be what's giving her those orgasms.

On the issue of penis size, then, men are continually getting mixed messages. Phalloplasty surgeons stoke male anxieties with advertising campaigns at the back of fitness magazines that proclaim "SIZE DOES MATTER," and humor abounds with jokes about the little member's sexual inadequacy (Q: Why are women such bad mathematicians? A: Because for years they've been told that this [thumb and forefinger a few inches apart] is eight inches. Q: What are the three most ego-deflating words a man can hear? A: "Is it in?"). In the HBO comedy series *Sex and the City*, four girlfriends lament and giggle over the "gherkins" and "miniature pencils" of past lovers. (It's not just men "behaving badly" on television nowadays.) But then, when a man seeks reassurance from his family doctor, he's told that size is irrelevant to a partner's pleasure.

Doctors routinely dispense reassuring bromides, sometimes drawing on biology. That's how Morgentaler goes on to justify his leg/penis analogy: "For this is the essential biological function of a penis in the first place, to enter the vagina so that semen is deposited near the opening of the uterus, called the cervix. A larger penis does not necessarily perform this task better than a smaller penis." Well, perhaps—if all a woman's wanting is to get pregnant. But even in that case, looking to biology—unless one does so in a highly selective way—will not necessarily yield reassurance for males. Sexual selection among many species is based on the principle of sexual dimorphism—females go for those males whose distinctively male traits (traits that the females lack) are bigger or more colorful than other males'. Darwin was the first to notice this, and to speculate that it was a kind of conspicuous "advertising" (he didn't use that word, of course) for prospective mates: "It cannot be supposed that male birds of paradise or peacocks, for instance, should take so much pain in erecting, spreading, and vibrating their beautiful plumes before the female for no purpose."

Sexual displays among animals are not just to attract potential mates of the opposite sex; they're also to assert dominance over potential rivals of the same sex. Interesting, isn't it, that this is just what those phalloplasty doctors promise to deliver to the augmented male: "Increase sexual confidence" with women. "Act like a man with a big penis" with other men. But strikingly, when it comes time for the biologists to produce examples of human sexual enticements and advertisements, the penis is nowhere to be found. Darwin alludes vaguely to "handsomeness" and "appearance." Popular science writer Deborah Blum produces the rather ethnocentric illustration of height. "Tall men," she writes, "are routinely rated most attractive by women, and in fact are more appreciated by all of society." (Does this hold true among Asians? I wonder. Or Eskimos?) James Gould and Carol Grant Gould have written a fascinating book called *Sexual Selection: Mate Choice and Courtship in Nature.* Although we learn a great deal about long tail feathers, big antlers, and enlarged claws in this book, there isn't a single entry in their index for "penis."

It isn't unusual for male primate display—which often involves males "flashing" their genitals to females—to be underplayed, even ignored, in such accounts—not because science has established that the penis is irrelevant to primate behavior, but because (I believe) we are much more comfortable with (and used to) telling the story in a way that emphasizes male activity and avoids putting the male in the passive position of having his genitals on display for "selection." But scientists and scholars can be remarkably obtuse about the ideological assumptions and biases that affect their work. Commenting on the use of penis pins among the Dyaks of Borneo, anthropologist D. E. Brown dismisses the notion that such pins (called palangs) could bring pleasure to women. "The neurology, physiology, and anatomy of the female genitalia provide little or no clear evidence [of this]," he states decisively, adding that according to Kinsey and his associates, "the inner walls of the vagina are generally insensitive." I've read some version of this finding over and over in the medical literature on phalloplasty. Even Thomas Laquer, a proponent of the social constructionist view that the way we interpret sexual data is always mediated by ideas about gender, is remarkably confident that he's describing objective "fact" when he writes that the vagina is a "far duller organ" than the clitoris. Elsewhere, he describes the vagina as "impoverished."

Oh really? Reading these "scientific" findings, I was reminded of the arguments of some early "second wave" feminists who insisted that vaginal orgasm is a myth concocted by Freud and other male scientists intent on yoking female sexuality to intercourse and reproduction. I liked the critique of ideology, but I didn't like the fact that I was asked to pay for it by denying my own experience. Like those feminists who claimed that only something called a "clitoral orgasm" was real, Brown and his colleagues' construction of reality ignored the testimony of many actual women. "Before Paul got his palang, I thought I had a good sex life," says one Dyak woman. "Now I tell him if he ever takes it out, it constitutes grounds for divorce." "Much more exciting than just a regular penis," says another.

Are these women having tactile hallucinations? More to the point, does it matter? Perhaps the physiology of the vagina is different than had been supposed (as some now argue, pointing to the tremendous amount of clasping and other vaginal activity that takes place during orgasm). But perhaps *ideas* have something to do with the pleasure the palang gives too. The rhinoceros, revered by the Borneans for its vitality, virility, and fertility, has a "natural" palang. In altering his body to be more like that of a virile animal, does the Bornean man enhance the fantasy power of his penis, and thus his ability to stimulate and excite his partner?

We don't have to choose between physiology and fantasy. In fact, we do so at peril of radically misunderstanding the *kind* of physiology we have—one that is tremendously suggestible to the cultural "superstructure" of ideas, associations, images. Whatever our nerve endings, sexual excitement and orgasm are never simply a matter of touching the right buttons. Or perhaps better put, not all our "buttons" are so clearly marked and located as the clitoris. Some of them can't bear to be touched at all, and thrive on distance and denial. Some of them are so diffuse that only the merger of two bodies will satisfy them. Some of them are in our metaphors: throbbing members, hot honeypots—or, on the nastier end of the spectrum, drills and slits. "Huge, throbbing members" may send some people into ecstasy, while "big dicks" will repulse them (or vice versa). A lot may depend on just *what* you imagine you have inside you. We could thus poll a hundred people, gay and straight, and probably get a pretty wide range of answers to the question of whether or not size matters to pleasure. Some preferences would undoubtedly reflect differences in anatomy and physiology. None of them, however, would be describing some "purely" physiological or anatomical set of facts.

Not that physiology is irrelevant, or infinitely mutable. But penises, like breasts, are heavily invested with symbolic meanings of different sorts. Even as sophisticated an evolutionary biologist as Jared Diamond utterly misses this symbolic dimension, as he ponders the mysterious evolution of the human penis to fourfold its size over the last 7 to 9 million years. Gorillas measure 1.25 inches on an average erection and, as Deborah Blum drolly puts it, "still manage to get a female gorilla pregnant. So why, exactly, does a human male need five inches or greater?" Diamond briefly considers, then basically discards the notion that the size of the human penis has evolved to advertise potency to prospective mates. I don't want to—don't feel qualified to—evaluate this as an evolutionary hypothesis. What interests me, rather, is the way that Diamond argues against it, citing the fact that women "tend" to report that "the sight of a penis is, if anything, unattractive. The ones really fascinated by the penis and its dimensions are men. In the showers in men's locker rooms, men routinely size up each other's endowment."

Diamond is right, as we've seen, about men's interest in comparative size. But what he misses, whether in connection with men's interest *or* women's, is the fact that finding a large penis (or any penis) visually attractive is not required to revere its potency or virility. (Among the Borneans, we should note, the well-hung rhinoceros is not renowned for its beauty.) In the days when I shuddered at the thought of looking square at a penis, I was well aware that the penis was an organ of formidable power. At a slightly younger age, I had cringed too over the pictures of the beast in my illustrated book of fairy tales. But when he put his huge paw on Beauty, I was thrilled (an erotic subtext that Disney's recent cartoon exploits to the hilt with its highly sexualized beast). Beauty is *not* an essential feature of the beast, but of the girl who liberates the prince in him. While movies may poke fun at the homely, nebbishy nerd, unhandsome but "virile" male actors like Telly Savalas and Dennis Franz have frequently become sex symbols in our culture.

My point is not so much about beauty, though, as about ideas. It may turn out, as some scientists claim, that we are drawn to each other simply through smell. If so, we have constructed an elaborate edifice of ideas to obscure that fact. It is ideas that endow the body with beauty and ugliness, ideas—if one subscribes to an evolutionary view—which advertise our dominance or reproductive "fitness" (or lack of it) to each other. Full lips, sociobiologists like to point out, are favored in women because they signify a high estrogen endowment. Perhaps. But the preference—as seems not to be the case for the male dog who will run blindly across several fields in hot pursuit of an odor—is one which culture can invalidate too. Consider the aesthetic racism that prevailed in the West until very recently against black and Jewish features; was that not perverse, evolutionarily speaking? Equally perverse (from an evolutionary point of view) seems the current cultural preference for skinny or muscled female bodies; if such bodies advertise anything, it's reproductive inadequacy, since female bodies require a certain level of body fat in order to produce the estrogen required to sustain reproductive cycling. Is aesthetic preference for such bodies a mode of "natural" population control? If so, it would be a strangely selective one, affecting those portions of the world's population—relatively affluent groups in advanced, industrialized countries—most likely to use birth control anyway.

We are creatures of biology *and* creatures of the imagination. I was struck by *Psychology Today*'s finding that women who rate themselves as highly attractive were much more concerned about penis size than other women. "Of women describing themselves as 'much more attractive than average,' " psychiatrist Michael Pertschuk reports, "64 per cent cared strongly or moderately about penis width and 54 per cent cared about penis length. Women who rated their own looks as average were about 20 per cent lower." The large penis, then, may be as much a status symbol, proof of entitlement to the best that nature has to offer, as it is a pleasure wand. Which isn't to say that large penises (relative, of course, to their partner's size) don't have a potential for contact and stimulation that smaller penises may lack, and this potential may indeed explain why the human penis has evolved to its present size. But it's also true that such stimulation, unavoidably, needs to be "interpreted" as pleasurable or unpleasurable by the person who is feeling it, and as human cultures have developed, they have provided a great many associations, images, ideas, to assist the body in this task. Does size matter? Absolutely, yes. But the matter of size is as "mental" as it is "material"—never just a question of nerve endings, always a collaboration with the imagination, and therefore with culture.

what is a phallus?

phallus worship and phallic symbols

"Always a collaboration with the imagination, and therefore with culture."

This phrase, with which I ended the preceding chapter, is also the key to understanding something about the penis that makes it unique among bodily organs—at least within a Western cultural context. From the Egyptian myth of Isis and Osiris to twentieth-century automobile advertisements, the penis has a symbolic "double" that is entirely the creation of the cultural imagination: the phallus.

The phallus is an extraordinary construct, and quite a slippery one to lay hold of. Let me start by comparing and contrasting the penis to the heart—another organ that has a symbolic double, although one that is not nearly so mysterious and complex as the phallus. There's the physical organ that pumps blood through the body and there's that quality—"heart"—that, while based on an analogy with the physical organ, has an extensive and varied cultural life of its own, from cartoon representations on valentines to song lyrics and everyday jargon. So too for the penis, although we don't have a common cultural recognition—the way we do with pictures of the heart's double—of what counts as a representation of the phallus. The pictorial symbol for "heart" is conventional; although this ♡ is not a particularly accurate representation of what a "real" heart looks like, we all know how to "read" the cultural symbol. In contrast, while some people may see the phallus in the design of a rocket, or in Joe Camel, others would find that idea laughable.

Interpretation is necessary because the phallus's anatomical double—the penis—has generally been taboo within Western culture, and so the phallus usually appears in veiled or abstract form, its relation to the penis obscured. Sometimes, those who create the symbols are not even aware of what they are doing (or at least that's the premise of much psychoanalytical and feminist theory). This wasn't always the case, however, and it's starting to change again today. Egyptian worshippers of Osiris and Greek worshippers of Dionysus carried unmistakable (if not anatomically detailed) representations of erect penises in their processions—and they knew very well that they were engaging in phallus worship. In the second half of the twentieth century, fashion designers have brought anatomically explicit phallus worship back into the public domain, with advertisements whose focal point is the model's large, often suggestively erect penis. However, contemporary viewers of these ads may not be as aware as the Egyptians and Greeks were of the cultural meanings attached to the phallic member, or that these advertisements are encouraging them to worship at its shrine.

Later strains of Greek culture pushed "literal" forms of phallus worship into the closet, and until the recent emergence of phallic underwear models, we have dealt mostly in what psychoanalysts call "phallic symbols," which draw (whether consciously or unconsciously) on analogies with the erect penis to make their point. So, in numerous ads, the "bodies" of automobiles arguably function as phallic symbols, as in the following two illustrations. What makes the 1946 Chevy (a relic of a far less sophisticated Madison Avenue) a phallic symbol? Not *just* the fact that it is "the longest of the lot," measuring "a big, strapping" fifteen feet plus. Those details may establish the analogy between the car and the humongous, erect *penis* (today we are a bit more subtle, alluding in one ad to "body-stiffening rail-through construction"), but they don't yet make the car into a *phallic* symbol. For that, the length of the Chevy must be "thrilling," making it "every inch the king" of cars. It's not just the visual or verbal allusion to penislike anatomical features that makes a car (or a rocket) a phallic symbol (those allusions may in fact be pretty schematic or obscure); rather, the suggestion of masculine authority and power is required. So, that thrusting Toronado can be handled only by the "all-man

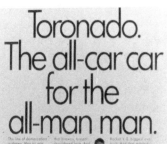

Cars that separate the men from the boys

man" who knows what to do with it. In the copy of this ad, the car-as-penis symbolizes not just the organ but the entirety of the "masterful" male body: "brawny, broad-shouldered," on the side of the men, not the boys. A car with *muscle*, both literally and figuratively.

The phallus is the penis that takes one's breath away—not merely because of length or thickness (qualities that might be sexually exciting but not necessarily command respect, as we've seen) but because of its *majesty*. Those who gaze upon it immediately feel themselves to be its subjects. That's the way phallus worship began, in ancient Egypt. In the Osiris myth, which tells the origins of such worship, the god Osiris is killed and dismembered by his enemy Typhon, who scatters the pieces of his body in all directions. Osiris's faithful companion Isis retrieves all the parts of Osiris except one—his penis—which she replaces with a replica, which she orders all to worship as a symbol of Osiris. So, in ancient Egypt, phallic icons were worshipped in temples, carried in processions, and so on, practices copied by the Greeks in their worship of Dionysus.

The Greeks, as I noted in the preceding chapter, also represented satyrs and other less exalted, sometimes comic figures with large, erect penises. Many scholars use the term "phallus" to apply to both sorts of representations. I would argue, against that practice, that caricature and comedy negate the penis's potential—whatever its size—to function as a phallus. To be a phallus, *reverence* of the erect penis (or its analogue) is required. Long Dong Silver, as we've seen, is *not* a phallic image; neither were the satyrs in ancient Greece. The element of special respect is present even in the old cultural habit (before theorists got their hands on the term) of using "phallus" simply as a fancy, more dignified synonym for penis. Not every penis, this practice suggested, is worthy of the term "phallus."

At this point, we can begin to see the way in which the phallus, although a symbol of the penis, is both similar to *and* different from the other symbolic constructs that I've mentioned in this book, like our literary penile metaphors—the "throbbing member," "engorged manhood," and so on. I called these "biometaphors" and suggested that they are among *our* verbal counterparts to the physical hyperboles (longest tail, biggest horns, brightest colors, decorated penis) of other species. What makes these animal hyperboles "metaphorical" is that they are generally not *functionally* of all that much use, but rather act purely as an advertisement of superiority. In 1982, Swedish geneticist Malte Andersson did an experiment that proves this point. Male widow birds drag around enormous tails of six feet or more in length. Andersson glued extra feathers on some, snipped some short, and left others as they were. The augmented tails won out (tails over head, as it were) with the females, attracting more mates than even the "showiest natural tail" and making it virtually impossible for the snipped birds to find mates.

Big, heavy tails—like the enormous antlers of many species of deer and elk—are not just useless, but a downright impediment to defense against predators. They make moving through vegetation more difficult and flight more cumbersome. Big, colorful crests can attract a predator's attention just as easily as that of a prospective mate. With the exception of those features (antlers, for example) used in ritual, rarely harmful contests with other males, the sole value of these sexual dimorphisms is to advertise the overall genetic superiority of the animal who displays them.* Similarly, the guy who is intimidated by the large penis of the cocksman with the locker next to his in the gym isn't afraid of what that burly penis could do to him in a fight; but he probably *does* tend to imagine that the guy who flaunts it so brazenly has the "balls" to clobber him. As far as sexual function goes, the large penis, as I've suggested, may indeed be capable of giving more pleasure than smaller penises, but not reliably so, given the powerful role that individual history and cultural meaning play in the human experience of our bodies; nonetheless, the large penis remains a powerful symbol of male sexual potency.

Unlike other animals, we often use words to excite and stimulate (and raise dread in) the imagination, hence our stock of vivid metaphors for sexual organs. But, as ads for muscle-building products and the booming business in penile augmentation illustrate, we use our bodies to advertise masculinity too. Unless one is a manual laborer, muscles have little use value in our management- and service-oriented culture; the potency of muscles resides largely in their cultural mean-

---

*Evolutionary theorists do not all agree on whether that advertising is "truthful" or not. Some argue that since the dimorphisms may hinder flight or fight, the animal is "lying"—he's actually less equipped for survival. Others claim that the huge energy investment in useless features could only be borne by an animal of superior endowments in other respects.

ings. Those are varied, but the display of masculinity remains chief among them. How else to explain why the ideal male body (as depicted in the underwear ads and the like) has become more ostentatiously muscular as women have started to pump up too? Clearly some of us, at least, want to retain those dimorphisms!

I have spent some time on explaining the concept of a "biometaphor" because one way of understanding the symbolism of the phallus is by way of its continuities *and* its differences from these biometaphors. Like them, the phallus stands for a superiority that is distinctively connected with maleness. But unlike them, the phallus stands, not for the superior fitness of an individual male over other men, but for *generic* male superiority—not only over females but also over other species. And unlike the biometaphors, the phallus stands for a superiority that is not just biological, but partakes of an authority beyond (and often in contest with) the power, needs, desires of the body. The biometaphors symbolize qualities (such as sexual or reproductive potency, superior aggressiveness, the capacity to give pleasure) whose value is not unique to human cultures but is shared by many other species. The phallus, on the other hand, proclaims its kinship with higher values—with the values of "civilization" rather than "nature," with the Man who is made in God's image, not *Homo sapiens*, the human primate.

I have been speaking of "the phallus" as if it were a timeless construct. In fact, the phallus as I have described it developed over time. The phallus actually started its cultural career with much more in common with the biometaphors of other species than it has now. Originally, like other biometaphors, the phallus advertised the superiority of an individual male—for example, the god Osiris—and in terms of "bio" qualities such as virility and generativity. The phallic images that Osiris worshippers carried in their processions, scholars agree, were signifiers of "rejuvenation and sexual pleasure" and—as Voltaire interpreted them—human gratitude to the gods "for making the human frame so instrumental in the perpetuation of the human species." These ancient meanings of the phallus linger in dictionary definitions; the phallus, as defined by the American Heritage Dictionary, is "a representation of the penis and testes as an embodiment of generative power."

Even in the worship of Osiris, however, a new element had come into the picture that took the phallus significantly beyond the scope of a more banal biometaphor. Osiris was not some member of a band of ordinary males, scrambling for position and mates, but a god. Thus entered an element of divinity absent from other biometaphors. And although Osiris (and his Greek counterpart, Dionysus) may have been the equivalent of male fertility gods, Western culture was to move rather decisively away from that "biological" strain of religious worship and toward those Egyptian traditions and their Greek counterparts which emphasized individual purification and salvation in the afterlife rather than the cyclings of the natural order. In the older traditions, the biological cycles of death and generation were a kind of species "immortality," a process in which the sexual, reproductive body played an essential role. In the new strains, the individual's hopes for immortality depended on achieving personal salvation, increasingly construed as a process of rising above the needs and desires of the "animal" body. Thus, sexuality ceased being a divine instrument of pleasure and procreation, and became a problem for the soul (as it was, for example, for Augustine, whose soul was rent in two by his sexual needs and his love of God).

As this happened, it became harder for the explicit depiction of the large, erect penis to serve as a symbol of anything divine or worthy of reverence. Instead, it became relegated to the symbolization of excessive, "animal" instinct (on satyrs, for example, as noted earlier), while the worship of the male principle became more abstract and its old, explicit penile reference more obscure. This worship began to focus too on those qualities that man fancies distinguish him not only from women but also from the other, "brutish" animals, creatures of pure instinct. Increasingly, masculine power and authority were seen to derive not from nature, but from God—not a god of procreation like Osiris, but the God of Immortal Forms, Disembodied Spirit, Pure Reason. Not surprisingly, the penis began to be seen as an object of shame, a rebellious little piece of flesh that kept pursuing the body's irrational desires. Kant, in fact, located the very origin of reason in the use of the fig leaf to conceal the male organ! Thus notions of male superiority ceased being grounded in sexual potency, and began to be grounded in the superiority of male intellect, rationality, mind—qualities that advertised not the superior virility of an individual male or the sexual generativity of the race but the divinely bestowed fitness of *men* to rule the earth.

"The phallic principle that gives man dignity and is worthy of worship is his ability to rise to the occasion, to answer the call of history," writes Sam Keen, in his typically turgid prose. Like the penis, the phallus must be able to rise to the occasion—but in the conscious service of history, not blind animal instinct. Remember the steel-hard, upward-pointing bodies of those two stalwart Nazi "comrades"? Their penises are unremarkable; but their torsos perfectly exemplify the authority and power that the possession of a penis (the right kind of penis, of course; a Jewish penis wouldn't do at all) is supposed to confer on its bearer. In this statue, the penis's potential for "hardness" and all that it suggests has been displaced onto the whole male body, where it can function more unambiguously as a symbol of strength, power, and upward aspiration. Where it will suggest Prometheus, not Priapus.

The example illustrates not only the male body depicted so as to suggest masculine authority and power, but the use of analogies with the erect penis to do so. But "hardness" alone is not sufficient to make a phallus, and well-defined muscles are not enough to make a master of the universe. The master body must signify an alliance with the gods rather than the masses, the heavens rather than the earth. He gazes outward and upward, undistracted, penetrating some higher truth.

As we saw with the Chevy and Toronado ads, it's not just male bodies that are "analogized" in this way. Penises themselves may be hidden away in our culture, but phallic symbolism is arguably *everywhere*. Remember the young guys' names for penises: masterful warriors and mythical heroes, torpedoes, rods, and rockets. When grown-up men design majestic skyscrapers and elongated rockets, do they have the erect penis in (unconscious) mind? We can argue endlessly about that. The proof, I would claim, is not so much in the suggestive form of objects, but in the metaphors we use to describe them. Conversations among scientists and defense intellectuals are so loaded with penile metaphors for their bombs and rockets that they would be hilarious—if they weren't describing events that decimated whole communities of people. "Then, just when it appeared as though the thing had settled down," reads one description of the explosion at Nagasaki, "there came shooting out of the top a giant mushroom that increased the size of the pillar to a total of 45,000 feet. The mushroom top was even more alive than the pillar, seething and boiling in a white fury of creamy foam, sizzling upward and then descending downward, a thousand geysers rolled into one."

Ads for military weapons, as Carol Cohn remarks, "rival *Playboy* as a catalog of men's sexual anxieties and fantasies." She illustrates her claim with an advertisement for the AV-8B Harrier II (don't ask me what that is) from a June 1985 issue of *Air Force Magazine*:

> Emblazoned in bold letters across the top . . . "Speak Softly and Carry a Big Stick." The copy below boasts "an exceptional thrust to weight ratio" and "vectored thrust capability that make the . . . unique rapid response possible." Then, just in case we've failed to get the message, the last reminds us, "Just the sort of 'Big Stick' Teddy Roosevelt had in mind way back in 1901."

Cohn doesn't remark on it, but "big stick" Teddy Roosevelt was one of the prime architects, at the turn of the century, of a new, "virile" conception of masculinity meant to save men from the feminizing and softening effects of excessive civilization.

What counts as phallic symbolism is always a matter of interpretation. When I first present arguably phallic images to my undergrads, they tend to react as Freud did when his theories were turned on his own personal habits: "Sometimes," he said, "a cigar is just a cigar." Of course he was right. But let's not be willfully obtuse either. Can there be any doubt that the White Owls in the ad on the facing page are more than "just" cigars? Perhaps in order to convince my undergrads I'd supplement the image with some of the numerous cigar-ad texts which associate cigar smoking with masculine authority: "I walk into a restaurant a little unsure of just how to hold my Corina Cigar as I move to my table. Several women sneak glances at me. The waiter seems more respectful today. I think of Albright at the office calling me 'Sir' this morning. I'm seated and I take a puff. I order . . . seems to me my voice is a little deeper and stronger . . ."

That's from a 1959 ad. The ideas, however, are still culturally current. As a symbol of masculinity, the cigar, in fact, is having something of a cultural renaissance today—and the associations are unambiguous, if sometimes playful. In Ridley Scott's *G.I. Jane*, the head honcho of the Navy Seals base where Jordan O'Neill (Demi Moore) wants to

prove her mettle like a man lights up his "fine $3.59 cent cigar" and goads Jordan, wanting to know if "the phallic nature of it happens to offend your fragile sensibilities." (The base captain, clearly, hadn't gotten a good look at Demi's neck muscles.) Later in the film, the Secretary of Defense himself, playing pool, lights one up, not because he likes to smoke but in order "to chew the hell out of it." With less sense of irony, the heroes of *Independence Day* light up after they have penetrated the alien stronghold, in what is clearly (and nonparodically) presented as a ritual celebration of manly accomplishment.

I gave Jordan O'Neill a resounding hand when she tartly answered the base captain, who inquired about her sensitivity to phallic symbolism: "The shape doesn't bother me, sir, just the goddamn sweet stench." As someone who grew up surrounded by clouds of my father's acrid, suffocating cigar smoke, I am better aware than most of the (usually unacknowledged) aggressiveness of the cigar-smoking king. Yet when I was living in Canada, I smuggled Havanas across the

Canadian border for my father. It was a gesture of love, but also of fealty. I knew that my father's cigars were more than just cigars.

Issues of interpretation get really complex when the most abstract and attenuated forms of the phallus are at stake, such as the "phallogocentrism" that deconstructionists and feminists have claimed runs throughout Western philosophy, science, and religion, or the Lacanian "privileged signifier," which is completely disconnected from the penis (or so Lacan argues), a disembodied, genderless construct which stands for abstractions like "the rule of law" and "the symbolic order." I have avoided dealing with these kinds of arguments here (although I have made them myself in other contexts) precisely because they move discussion of the phallus so extremely far afield from the male body, and I fear they will pull me into an abstract realm where I would be vulnerable to the very disease that I would be diagnosing.

I certainly agree with Lacan that the phallus belongs to the realm of ideas, not biology; it's a symbol, not a body part. But the symbol emerged historically, as I've suggested, out of forms of reverence that *did* have reference to biology, and these references themselves have analogues in the morphology and behavior of other male animals. The meanings that we have attached to the phallus, moreover, no matter how abstract and attenuated they have become since the time of the early phallus worshippers, are grounded in the bodily image of the erect penis and ideas humans have attached to *it*. From this perspective, to proclaim that the phallus has nothing to do with the penis is to suffer from a sort of advanced "phallus" complex oneself, in which mind stands supreme over body, human over animal, symbol over flesh. Poised here as we are on the brink of the most abstract reaches of the phalluses, let's return instead to the penis, and the role the phallus plays in the bodily life of real people.

fictional phalluses

Just as a cigar is sometimes more than a cigar, sometimes the penis is more than just a penis too.

The phallus is a cultural icon which men are taught to aspire to. They cannot succeed. Young men, you will recall, have trouble seeing their penises realistically, and consistently judge them to be smaller than they actually are. In part, that's because it's not really flesh-and-blood penises that shape a young man's perception that his penis is less than it is or should be, but a majestic imaginary member, against which no man's penis can ever measure up. As psychologist Anthony Quaglieri insightfully notes, thinking that one's penis is smaller than it should be is not really about inches but "about how men are trained by the world to see ourselves as not enough." Even those who *think* that they measure up in size, cannot measure up in heroic, unflinching constancy. The phallus, we might say, haunts the penis. Paradoxically, at the same time the penis—capable of being soft as well as hard, helpless as well as proud, emotionally needy as well as a masterful sexual performer—also haunts phallic authority, threatens its undoing. Of course, I'm no longer talking about rockets or automobiles. I'm talking about those cultural images that depict the penis as though it were a phallus, and in doing so hold manhood hostage to an impossible ideal.

The penis becomes a phallus when, like the "thrilling" kingly Chevy, it is viewed as an object of reverence or awe, as in this famous passage—again a cultural prototype—from *Lady Chatterley's Lover*. Earlier, we looked at Connie's description of Mellors's "soft" penis. In the following description, it is transfigured, not merely into a "hard" penis but into a phallus:

He dropped the shirt and stood still, looking towards her. The sun through the low window sent a beam that lit up his thighs and slim belly, and the erect phallus rising darkish and hot-looking from the little cloud of vivid gold-red hair. She was startled and afraid.

"How strange!" she said slowly. "How strange he stands there! So big! and so dark and cocksure! Is he like that?"

The man looked down the front of his slender white body, and laughed. Between the slim breasts the hair was dark, almost black. But at the root of the belly, where the phallus rose thick and arching, it was gold-red, vivid in a little cloud.

"So proud!" she murmured, uneasy. "And so lordly! Now I know why men are so overbearing. But he's lovely, *really*. Like another being! A bit terrifying! But lovely really! And he comes to *me*!" She caught her lower lip between her teeth, in fear and excitement.

The man looked down in silence at his tense phallus, that did not change. . . . "Cunt, that's what tha'rt after. Tell lady Jane tha' wants cunt. John Thomas, an' the cunt o' lady Jane."

These passages were my earliest form of sex education. I had an older sister in college, and she had all the forbidden books on her bookshelf. I would sneak in while she was off at school and read *Lady Chatterley* over and over again, panting. Looking at these passages again now, it is no wonder to me that I grew up with such a mixture of reverence and fear of the male organ—feelings that were present for me even in the experience of petting with Bobby Cohen—awakened, significantly, when I touched his hard penis. Bobby Cohen, lying on top of me, humping away, was just a little boy masturbating with my body. But when I felt his hard penis, he briefly joined the company of Mellors and all those proud phalluses that had fired my imagination in my sister's bedroom. Lovely. Terrifying. Lordly. (To quote Connie Chatterley.) And he comes to *me*! For Bobby himself, I am sure that the experience was not quite so archetypal. Bobby, an adolescent boy groping his way to manhood, would have been amazed (and probably terrified) to discover what was beginning to simmer in my psyche.

The phallus, we must always keep in mind, is an idea, not a body part. I imagine that as far as Bobby was concerned, what I was touching was not a phallus but a hard-on—and one that neither of us knew quite how to handle. In contrast, during the era in which Bobby and I came into adolescence and young adulthood, many literary cocksmen often arrogantly conflated *their* hard-ons with earth-moving phalluses. In "The Time of Her Time," Mailer's alter ego Irish narrator Sergius O'Shaughnessy (who elsewhere describes himself as "a Village stickman who could muster the divine It on the head of his will") explains his awakening "desire" for a young Jewish girl (why do we always have to play foil for these guys?) he's met at a bar:

> Her college-girl snobbery, the pity for me of eighty-five other honey-pots of the Village aesthetic whose smell I knew all too well, so inflamed the avenger of my crotch, that I wanted to prong her then and there, right on the floor of the party, I was a primitive for a prime minute, a gorged gouge of a working-class phallus, eager to ram into all her nasty little tensions. I had the message again, I was one of the millions on the bottom who had the muscles to move the sex which kept the world alive, and I would grind it into her, the healthy hearty inches and the sweat of the cost of acquired culture when you started low and you wanted to go high.

I won't "unpack" everything that is going on this passage—by way of anti-Semitism, myths of female power, romanticization of the working-class hero, and so on. In 1997, it all seems embarrassingly obvious. But such passages *didn't* strike me in that way in 1963—the year a (male) high school English teacher recommended the story to me for my enlightenment. The era was fully at one, for the time being, with Mailer's macho version of sexual liberation, and so was I. I found "The Time of Her Time" exhilarating, raw, honest. I was completely oblivious to Mailer's posturing and bullying, and even less to his sexism. But on the other hand, I was primed—by my male teachers, the male book reviewers, my boyfriends, all swept up in the male rebel ethos that was so prevalent then among intellectuals—to see this as transgressive and liberatory literature. I had no other paradigms for understanding it. Rereading some of this stuff today, I am amazed at the lessons in sexuality, gender, and race that I was swallowing, along with the throbbing prose.

Mailer is the poster boy for an especially self-satisfied and violent penile narcissism masquerading as a "working-class phallus" on a mission to cure civilization of its discontents. "A stiff prick has no conscience," he says in "The Time of Her Time." For Mailer, this represents nature's way, which should always be obeyed against the demands of "civilization" and its feminizing, domesticating inclinations. When the prick is hard, nothing should stand in its way. When it's soft, that's because some woman has interfered with the cosmic order (by using birth control, for example). Another, more contemporary spokesman for a phallic "cosmos" in which women are put in their rightful place is Sam Keen. In one passage of his book *A Fire in the Belly*, he brags about how a female member of the audience paid homage to one of his "Understanding Men" seminars. "This weekend I feel like I have been in a room with giants," Keen proudly reports her saying. "I thank you for letting me listen."

In another passage, which I quoted earlier, he describes looking in the mirror at the end of a night of sex: "The mirror reflected the image of a man I had never seen before—his cock, resting but proud, pulsated with life, his chest swelled with the joy of being, his sinuous muscles were full of power." This description is presented utterly without irony or humor. More remarkably, it comes at the end of an anecdote

that is meant to demonstrate Keen's triumph over the sexual performance principle, on a night when he was "not ready" (his choice of
words) for sex despite his partner's arousal. She reassures him, but he
feels he has "failed the test of manhood" and drifts off "into a restless
sleep." Then, in the middle of the night he awakes, fully erect, and enters her ("and [his] manhood," as he tells us) with barely a word; and
they "moved together until the whirlpool sucked us down into the primal deep, and we slept as one body in the sweet darkness." Later, Keen
wakes up and "roars like a lion."

To be fair, the context of this anecdote is a quite perceptive critique
of "getting laid and keeping score" (Keen does make some good points
in the book). It is clear that *he* views the incident as having brought
him to a recognition that sex can never be deeply satisfying to men so
long as it is viewed as a test of potency and performance. I agree. The
anecdote, however, seems not to demonstrate this so much as Keen's
need to be the active, controlling, "deciding" partner in sex. It would
be one thing if Keen were to explore the psychological or cultural un-
derpinnings of this need. Instead, he is more interested in creating a
new masculine icon out of his own "transformation." His "new vision
of manliness" is bursting with phallic adrenaline and metaphors. The
call is to create a "new race of giants," "fierce gentlemen," with
"thunder and lightning in them," men with "potent doubt," "manly
grief," "virile fears." I know that Keen wants to *legitimate* male doubt,
grief, and fear here. But what can be the point of doing so through the
invocation of "virility" and "potency," if not to "pump up" the image
of the new man with a promise of phallic mastery? And if the new man
is a giant and master, then in what sense is he "new"?

But the penis and the phallus can collaborate in other ways than
through the restoration of male dominance. Gay male cultures are
hardly immune to fetishizing phallic power, as the preponderance of
phallic imagery and symbolism in homoerotic art testifies. But gay
male art and literature often also play in an ironic and self-aware way
with phallic icons, parodying and drooling over them at the same time.
In Tom of Finland's "Hitchhiker," for example, the symbols of aggressive masculinity—biker's tattoo, pirate's earring, gladiator's cuff, police
cuffs, frontal bulge—are so overabundant that they simply cannot be
taken seriously (as, for example, Mailer and Keen take their own
equipment so very seriously). At the same time, this is not just a car-

Ironic phalluses: the symbols of masculinity
are so overabundant they become playful
parodies of themselves ("Hitchhiker" courtesy of Tom of Finland Foundation)

toon; it's erotic in intent (folks masturbate to Tom of Finland's images)—but with an ironic wink. The same kind of sensibility is illustrated by Brian Pronger, quoting a gay gymnast, gushing over male muscles: "I don't know what it is but there is something about men's tits that just does a number on my brain! A man's chest is just irresistible!" Finding male pecs irresistible is to revere male power. But calling them "tits" is to undo that reverence (perhaps—or perhaps not, depending on how one sees things—at the expense of women's dignity).

A similar sense of irony and play with the icons of phallic masculinity is found in Pat Califia's description of putting on a dildo:

> First the dildo harness, which I made myself out of two leather straps . . . Once it's in place, I pull all the straps snug (Velcro makes them adjustable; I love technology), put the jock on and bend over to pull on my boots. I put on my black leather chaps. I shrug into my motorcycle jacket and zip it shut with nothing underneath it—another broken taboo, for a woman. But I don't feel like a woman anymore. The semiotics have shifted. For instance, my long hair doesn't mean femme; it means hippie biker. And there's my cock. I can grab it, stroke it, and it feels *good*. It looks good too, the bulge covered by tightly stretched white cotton, framed by black leather.

Sounds like when she's finished, she'll look just like Tom of Finland's "Hitchhiker"! Other lesbian accounts, not so cocky as Califia's, nonetheless emphasize that a Mailer-like sense of entitlement is not the only way to experience a phallic moment. Susie Bright quotes a woman named Lorraine:

> I never thought of myself with a phallus. I bought these toys quite naively, thinking it was fun and different. But I wasn't prepared for how different I felt fucking my girlfriend with a dick instead of my hands. At first I felt ridiculous. Then, after I actually started fucking her with it and she was responding, this wave of euphoria came over me, which was followed by an even bigger wave of shame. I've always thought of myself as androgynous, but that was in terms of a middle ground, not a woman who had sharp masculine desires or any trace of masculine identity. I had to pinch myself the next morning to see if it was still me.

Pat Califia provides some terms that may be helpful here. She distinguishes between the "male phallus" (what I would call, more accurately I believe, a "masculinist phallus") and the "butch phallus." The "masculinist phallus" is the phallus that Mailer wields. Its swagger, its toughness, its rough edges, its dominance is in the service only of itself; any satisfaction that the woman feels is purely an accidental break for her. "The Time of Her Time" describes O'Shaughnessy's battle (for that's what it is, and even Mailer admits it) to bring the female protagonist to her first orgasm. But the point of it all is to keep his reputation as a "stickman" intact; he's inflamed by many strong emotions—anger, competition, anti-Semitism, frustration, pride—but not once by the thought of giving the girl pleasure.

In contrast, the "butch phallus" offers its roughness, toughness, cockiness with a keen eye for when those qualities begin to interfere with rather than enhance a partner's pleasure. "If the butch phallus does not succeed in giving the femme sexual pleasure," Califia claims, "it does not exist." Similar notions are emphasized in many literary depictions and political discussions of the use of dildoes among lesbians. In Leslie Feinberg's *Stone Butch Blues*, an older, more experienced woman gives a young butch a dildo, instructing her solemnly: "You know, you could make a woman feel real good with this thing. Maybe better than she ever felt in her life. Or you could really hurt her, and remind her of all the ways she's ever been hurt in her life. You got to think about that every time you strap this on. Then you'll be a good lover." Feinberg's ideal of the "butch phallus" is not only attentive to the needs of the other (behaving, one might say, in a traditionally feminine manner) but is also femin*ist* in its desire to heal the wounds of sexism. The "butch phallus," of course, is an ideal; neither Califia nor Feinberg is claiming that it goes with the territory of strapping on a dildo.

The phallus, remember, is not a real body part. Having one or not requires the permission of culture and/or the exercise of attitude more than the possession of a particular kind of body. Having a penis (even one that is imagined to be extremely large) does not protect one from being divested of phallic authority, from being a "man manqué" (as Richard Majors describes the historical situation of the black man). At the same time, *not* having a penis—although it has historically been a monumental impediment—is not an insuperable obstacle to projecting phallic authority. Nowadays, a body and attitude that's been properly developed can go a long way in making up for biology's lack of coop-

Demi Moore develops her phallus in *G.I. Jane* (*Photofest*)

eration. In *G.I. Jane*, after being beaten and bloodied by her training officer, Jordan O'Neill—who has defiantly gone through men's training and testing to prove her equal mettle—is told by him to "seek life elsewhere." The comment takes her over the edge. "Suck my dick!" she replies. That woman thinks she has a phallus—and who would deny it? The last time the expression "suck my dick" was uttered in such a challenging way in a feminist movie, it came out of the mouth of a would-be rapist, about to be told where to shove it by the blast of Louise's gun in *Thelma & Louise*. Ridley Scott directed that one, too. A woman doesn't need the gun anymore, he seems to be saying in *G.I. Jane*, if she's got the dick. I was expecting not to like *G.I. Jane*, having heard about this scene before I saw the movie. Generally, I'm not a fan of "power feminist" models of liberation. But I have to say that I was impressed with Demi Moore's thoroughgoing transformation of her body and all it projects from a toned-femme-with-implants aesthetic to a butch look and style. It was in the shoulders, in the way she walked, in her pleasure as she shaves her head. Demi made me believe that

pleasure was real. Some women stand "as though" they have phalluses; claiming space with their legs and groins in a challenging and
confident way. I like that.

Wanting my students to explore the degree to which gender is a language made up of bodily codes, I had a "gender-bending" party for my
class. Everyone was instructed to either dress as a caricature of his or
her own gender, or dress against their own gender. Many of the cross-
dressing women later reported dramatic changes in the way they experienced themselves "as bodies"—their arms and legs moved through
space differently, they sat differently, they flirted differently. One—let's
call her Claire—wrote about the experience as "Jason," her male alter
ego:

> I'm usually very outgoing, very friendly. But as Jason, I was belligerent,
> crude, uncontrolled—not because all men are this way by any means, but be-
> cause looking like a male meant I could be rude if I wanted to be without
> facing the "punishment" for rudeness usually conferred on women. I sat on
> the couch and spread out—taking up more room than I ever would have felt
> was my due as a woman. I drank three beers, belched loudly, and proudly
> talked about seeing a man about a dog. I drank in the spoils of male privi-
> lege that night, which were afforded me because of a surface change. By
> marking my body, I had unmarked my femininity.

Claire—a heterosexual woman, and a quite "feminine" one under ordinary circumstances—clearly experienced her temporary masculinity
as liberation to enjoy what Pat Califia calls the "male phallus" (that's
G.I. Jane's "dick" too). *New York* magazine reporter E. Jean Carroll
likes dressing up in men's clothes for the same reason—"the high, wild
sense of *superiority* they give me—the ineffable feeling of being, now
and forever, the big Kee-Mo-Sah-Bee."

Similarly, in the case of a woman who had a fantasy penis, psychoanalyst George Awad describes the sense of power that it gave her:

> When she was convinced that she had a phallus, she felt omnipotent and im-
> mune. In one instance, for example, she intentionally parked her car in a "no
> parking" area and was enraged when the car was towed away. She was cer-
> tain that nothing was going to happen to her car because she had a penis.

Claire, Carroll, and Awad's patient have a certain *idea* of what phallic
power involves, of course, and it's one that seems deeply shaped by

their sense of the constraints of *femininity*. Claire's pleasure in having a male body is connected to feeling permission to do anything she wants, taking up space defiantly, belching loudly, not giving a shit, forgetting about all those limitations she experiences in her life as a woman. It's the same conception of masculinity that Dewar's scotch has in mind, when it illustrates the theme of "getting in touch with your masculine side" by showing a woman drinking milk out of a carton.

But there are other ways a woman might understand the phallic experience, as we have seen, than as permission to slobber, belch, or blow acrid smoke in someone's face—and the same applies to men. Lesbian erotics—thought by many heterosexual men (and women) to be "man-hating"—actually enable men to imagine a way of being phallic that includes humility, tenderness, and mutual recognition. In fact, something akin to the butch phallus is, arguably, the male phallus that is celebrated in the romance novels: the throbbing manhood who takes his pleasure from the pleasure of his femme. The difference, of course, is that in the romance novel these heterosexual butch/femme dynamics are presented as a pattern of nature, without consciousness that they are a stylized *cultural* language of interaction—emotional, gestural, postural, social. Thus the absence in the romance novel of the irony and playful humor that is offered by gay and lesbian cultures.

If our culture would permit it, a man, no less than a woman, might just as well aspire to a playful "butch phallus" as to a deadly serious masculinist one, his sense of manhood on the line with every thrust. I've known men who can play with that "butch phallus," who are able to indulge a woman's fantasies of being dominated, her moments of phallic worship, without turning into an engorged avenger. With greater permission to mix it up genderwise, more and more men—and women—may feel comfortable putting on (or taking in) the phallus on a temporary basis, strictly in the service of Eros. Remember, the phallus is a creature of the cultural imagination, not biology. No one is born with one; no one can claim "the real article." Some men may think they can by virtue of having penises, but they are mistaken. The next time you hear a young guy meanly spurting "Suck my dick!" remember that he—not only Demi—is pretending.

public images

fifties hollywood

THE REBEL MALE CRASHES THE WEDDING

Even more than most baby boomers, I grew up at the movies. It wasn't just Saturday matinees with other kids. There were plenty of those, but for me they were never the real thing. The matinees were for eating candy, giggling with my friends, they were mostly a social event. I was bored by the endless cartoons, by child-and-dog adventures, by slapstick comedy, by cowboys and Indians. The real thing, for me, was going to the grown-up movies with my family—in Newark, on the evenings after my father had returned from a business trip, on Sunday at Radio City Music Hall in New York City, in Philadelphia or Albany, where we occasionally accompanied my father on a trip. Everyone in my family liked going to the movies. But only my father was a true fanatic, and I inherited it from him. On one day trip to Philadelphia, after my father had gotten his meetings with brokers taken care of in the morning, he and I saw *The Nun's Story*, *The Horse Soldiers*, and a full-length comedy that I can no longer recall before we finally gave in to exhaustion. My father had taken only me along on this trip; he had minimal business to do, and we both knew that I was the one he could gorge on movies with. At least, that's what I proudly told myself (and my friends) about our solo date.

It was 1959. I was twelve. My parents did not distinguish between adult movies and fare suitable for children. We saw everything that my parents saw. I don't think it ever occurred to either of them to worry about what we were learning from the movies or what the images conveyed to us, whether we were frightened, or bored, or confused. It may be that few parents did in those days. Or perhaps my parents were unusual in their relaxed attitude. My father's bedtime stories to us—some taken from myth and classic literature, but others from O. Henry, Damon Runyon, and recollections of his own street youth—were often full of violence and sexual suggestion anyway. What was the difference between viewing the scowling gangsters, slutty heroines, bloodied bodies in our imaginations and viewing them on the screen? For my father, all imaginative genres (including his personal autobiography, which often was embellished with extra drama) had the same point—creating a world more vivid, more condensed with meaning, more exciting, more instructive than the "real" world, and offering a temporary escape from it.

The movies my parents took us to did sometimes disturb me. I was only eight when I saw *Picnic* (1955), but it managed to tangle up love, sex, exposure, and humiliation powerfully enough to upset me for weeks—although I hardly understood what I was upset about. When I was a bit older and clearer on what sexual intercourse was, certain movie scenes addled my brain so uncomfortably that when I watch these movies today, the precise mixture of terror and excitement they stirred up comes right back to me. Shirley Jones won an academy award for playing against type as a prostitute in *Elmer Gantry* (1960). In one scene, she describes how preacher-con man Gantry (Burt Lancaster) seduced her when she was a girl, behind a pew in her father's church: "He got to howlin' 'Repent, repent!' I got to moanin' 'Save me, save me!' First thing I know he rammed the fear of God into me so fast, I never heard my father's footsteps." *That* speech even knocked the socks off my salesman father, who loved quoting it. The open lustiness of it, and its slightly sacrilegious penis joke delighted him. *I* was excited—and then disturbed *because* I was excited—by the power imbalance between Lulu and Gantry. Inexperienced Lulu didn't have a chance against Gantry's confidence, potency, will.

Hitchcock's *Psycho* (1960) was the first movie I watched peeking out from between my fingers, prepared to blot it out at a moment's notice. I was so scared, I even conjured up sexual fantasies that had nothing to do with the movie to compete with the unbearable suspense I was feeling. The result was that I got hot as well as terrified. *Psycho* was my first experience of turning inward during a movie, of "leaving" the movie to enter a world other than the one it presented to me. That was upsetting in its own way; it made me feel alone, weird, split off from the world. Imagine feeling weird for splitting off from *Psycho*! But when I was in a movie theater, it *was* the world. Before *Psycho*, I had always been able to vanish into the movies, give myself to them so completely that nothing else existed. Even today, movies are the supremely safe place for me, an absolute refuge from the anxieties of real life. I enter movie theaters with confidence that any bad psychic chemistry of mine will be alchemized, evaporated, by the time I leave.

Almost everything that I've imagined about men, expected from men, feared about men, has either come directly from the movies or interacted intimately with ideas and images from the movies. But for someone who loves the movies as passionately as I do, and who has seen so many movies that it's difficult to find something to rent at the video store when the new releases are out, my *knowledge* of just how images of men and male bodies affected me has been—I now see—surprisingly inchoate. Perhaps I preferred to keep it that way, sensing that a certain amount of embarrassment and shame, perhaps some painful confrontation with old feelings and events, would be the result of becoming more conscious about such things. When I decided to write a book about the male body, however, I knew I would have to revisit the movies I grew up with, and try to bring into clearer focus the then-obscure but powerful messages about men, sexuality, masculinity that they communicated to me (and, I assume, to at least some other women of my generation), and to try to understand something too of what they may have communicated to men.

I knew it would be a daunting task, and one I could undertake only partially and selectively. What I wasn't prepared for, however, was how very different everything would start to look once I put masculinity and the male body in the foreground. I began simply wanting to retrieve and explore how movie images of men and masculinity affected at least *one* female viewer—me—at crucial moments in my development. But to do that, I discovered—to explore movies from the point of view of a female subject looking at male bodies—was not just to provide a "female" perspective, but to fundamentally challenge some central operating assumptions of both film and cultural history.

For example, as a student and critic of film (I was a voracious reader of film criticism and did a weekly film review for a student radio station in the mid-seventies), I had uncritically accepted the standard history of the "sex object" in Hollywood movies as all about women (with perhaps a paragraph on Valentino). In 1966, a well-received book by Alexander Walker called *Sex in the Movies* contained three sections: "The Goddesses" (Mae West, Dietrich, Garbo, Jean Harlow, Monroe, Elizabeth Taylor, and so on), "The Guardians" (about censorship), and "The Victims" (about the emasculated male in Italian and American sex comedies). Where was the chapter on the gods? On male sex symbols? Those (straight) male critics and teachers of mine were not the most reliable sources, perhaps, on the sex appeal of the male body. But when I took courses in film in the sixties and seventies, I didn't even notice these gaps in their understanding—even though I knew quite well that I had spent much of my adolescence fantasizing about the bodies of various male actors and characters in the movies. The critics and my teachers had the official knowledge; my experience belonged to the realm of the merely "personal."

Once my own dim bulb got a little brighter, I realized that one reason why I had been inclined to accept the picture as presented to me was that the movies that were the rage in the sixties and emblematic of the "sexual revolution" of that decade—*The Graduate, Butch Cassidy, Blowup, 8½, MASH, Easy Rider, Midnight Cowboy*—*were* fairly circumspect with the male body, even with their more glamorous stars. Sure, sometimes someone would take his shirt off, but the moment was not eroticized in the way I remembered Paul Newman's body, Burt Lancaster's body, and William Holden's body eroticized in the films of the fifties. Was this a species of false memory syndrome? Were my recollections of the images I saw as an adolescent steamed over by the hormonal changes I had been going through at the time? Or could it be that there had actually been more and better male "skin" in the *fifties*? If so, perhaps our narratives about the sexual repressiveness of that decade needed to be revised somewhat.

I was also fascinated to discover that Elia Kazan's film of Tennessee Williams's *A Streetcar Named Desire* and Vincente Minnelli's *Father of the Bride* are almost exact contemporaries, released in 1951 and 1950, respectively. (This was too early for me to have absorbed them as first-runs in the theater, although I was undoubtedly there, snoozing or struggling on my mother's lap.) If ever there were contrasting male types, the two Stanleys of these films—Stanley Banks (the father of the bride, played to perfection by Spencer Tracy) and Marlon Brando's electrifying Stanley Kowalski—are those types. Kowalski—the abusive but sexually vital king of his household. Banks—the gruff but indulgent "good provider" who ultimately comes round to every demand his wife and daughter make. I began to see the two Stanleys as competing prototypes for the souls of men in the fifties, and progenitors of images of masculinity that have endured through the nineties. Kowalski spawned, among other types, the violent and misogynist action hero who still rules the box office (although by now he's a comic-book parody of himself). Stanley Banks was the forerunner of the scores of put-upon hubbies and fiancés who incompetently wage—and inevitably lose—the "battle of the sexes" in what Alexander Walker calls the "matriarchal sex comedy" of the fifties and early sixties, as well as more positive father figures like Jim Anderson (Robert Young) and Cliff Huxtable (Bill Cosby).

Although belonging to different film genres, these prototypes are in constant conversation with each other about manliness and its relation to domesticity. The action heroes are in rebellion against everything that Stanley Banks represents. Yet the forerunners of today's action heroes initially emerged in dramas like *The Long Hot Summer*, in which manly independence and settling down with a good woman were not at all incompatible. It's only later—in the sixties—that the "vigorous male animal" (as I call the dreamboats—often played by Paul Newman—that sauntered into women's hearts in the fifties) morphs into that being, well known to us today, whose world (except for the quick lay, the dead girlfriend, or the barely visible wife) is almost exclusively that of car chases, earsplitting explosions, contests with other men. It was inevitable that this hero began to have less and less to offer romantically to female viewers (even when played by actors like Harrison Ford and Mel Gibson, who originally appealed strongly to female viewers). I'm entertained by a good action movie. But my heart rarely throbs in the dark for these guys anymore. I'm more likely to wait for the video, so I can do something else while I watch.

Cultural change, however, never marches unwaveringly down one avenue. Marlon Brando's portrayals of macho consistently maintained a highly self-conscious, critical edge (Brando himself claimed to "despise" Kowalski), and were also to inspire a generation of actors and directors—Jack Nicholson, Robert De Niro, Martin Scorsese, Brian De Palma—bent on "deconstructing" macho rather than playing it straight. Brando also brought an emotional expressiveness, a willingness to portray male *need*, helplessness, dependency, that helped to shape a very different kind of romantic male ideal than the violent action hero. Even with Kowalski—as brutally heterosexual as they come—Brando was "queering" masculinity in ways that James Dean and others would develop still further, and that have remained a constant undercurrent of macho masculinity.

father knows best . . . and he wants out

The fifties had begun in a spirit of relentless pro-domesticity, focused on reestablishing American family life, getting men and women back together again after the war in happy, homey consumer units. It wasn't hard. Men and women separated by the war were, of course, almost mad with desire to be reunited. My father's long, beautifully written love letters to my mother, available to me later in his brown leather "war scrapbook," were full of Keats, Shelley, and Byron, and an intense romantic longing for my mother that formed part of my imaginings of a wonderful (and mysteriously eradicated) relationship that they had shared "once upon a time," leaving only paper evidence like this behind. Still, I recognized the father that I knew—if not his relationship with my mother—in the vivid descriptions of the beach at Iwo Jima, the long excerpts from love sonnets, his use of a Cyrano de Bergerac device to express his own longing by invoking the poets. Another sailor, however, must have inserted other details. My father, who never owned a house in his life and whom it is simply impossible to imagine cutting a lawn, rhapsodized poetically about a "cottage lone and still" in which he, my mother, and my older sister (then an infant) would live.

During the war, ads frequently appeared in *Look* and *The Saturday Evening Post* encouraging such fantasies. "First you dream . . . of victory and a home. Then you plan . . . with war bonds. And *tomorrow* it will all come true!" The explicit injunction of this particular advertisement, which features a sketch of a soldier and his wife cuddling on a park bench, sketching a picture of a house on the ground in front of them, was to buy war bonds. But the not very subtle agenda was to get men and women fantasizing about the "peace-time homes" that are going to be "easier to own . . . and better to live in," with "comforts and conveniences now considered luxuries," thanks to the "new electric age" being ushered in by General Electric (the company behind the ad). Hollywood lent its celluloid to this fantasy. Sophisticated, playful Cary Grant settled down to build Mr. Blandings's picket-fenced dream house. Spencer Tracy ceased being the equal movie partner of a sassy, classy Katharine Hepburn and became a suburban dad. Myrna Loy hung up her sly sexual allure and put on a frilly apron.

It was a world of gruff, kindly male providers and kittenish women-children, a world where the daily working grind was men's domain and women frittered their time away figuring out ways to spend men's money decorating their houses (while a maid took care of all the cooking and cleaning, of course). The two spheres worked in perfect harmony (in the idealized Hollywood household, that is, where every man had a job that brought home enough cash to bankroll his wife's frivolous projects) and provided great opportunities for a booming consumer culture to exploit men and women's desire to fit into their appropriate gender niche with the right material accouterments. Women's magazines helped this effort by arguing that a woman's only true fulfillment was as a wife and mother; popular sociology urged men to take on the role of breadwinner as a necessary component of masculine maturity and "normal" sexuality. As Philip Roth recalls in *My Life as a Man*:

> A young college-educated bourgeois male of my generation who scoffed at the idea of marriage for himself, who would just as soon eat out of cans or in cafeterias, sweep his own floor, make his own bed, and come and go with no binding legal attachments . . . laid himself open to the charge of "immaturity," if not "latent" or blatant "homosexuality." Or he was just plain "selfish." Or was "frightened of responsibility." Or he could not "commit himself" (nice institutional phrase, that) to a "permanent relationship."

Father of the Bride (1950) and *Father's Little Dividend* (1951) are virtual courses for men in how to normalize oneself according to this scheme of things. They are also charming, entertaining films, full of droll verbal and visual humor. Spencer Tracy, as father of the bride Stanley Banks, and Joan Bennett, who plays his wife, Ellie, underplay with deadpan precision. Elizabeth Taylor, as daughter and bride Kay, is stunning to look at (if a bit annoying to listen to). It's as hard *not* to enjoy these movies as it was not to enjoy *The Cosby Show* on television. The Bankses, like the Huxtables (but unlike television's Andersons and Cleavers, who were the historically closer relatives), live their bourgeois lives with just enough irony and sarcasm to disarm the gag reflex.

Having said all that, let me try to rearm that reflex. At the start of the first film, Stanley enumerates—*sans* any discernible irony—the required ingredients for male success and happiness. He's a partner in a law firm, every day he takes the commuter train home to a peaceful house in the suburbs ("almost paid for" in the first movie, all paid for in the sequel), his pretty wife is at the door with a martini, their faithful black maid ("our Delilah") is "always the same." Stanley has two sons who, ha-ha, eat him out of house and home, and an adorable daughter who, sigh, thinks he's her hero. Flies in the ointment? In the first movie, his little kitten wants to leave him for another man, and he has to foot the bill for her wedding. In the second, she's making Stanley a grandfather. Just when the mortgage was all paid and you thought "the whole world of adventure was waiting" for you (manly exploits like "fishing, game hunting, even climbing Mount Everest") you're expected to push a pram.

Women watching these movies today might be most offended, as I once was, by the portrayal of our sex. Especially in *Father's Little Dividend*, we hear the constant babble of female voices chattering, simpering, weeping, fussing over wallpaper, house plans, whether or not their husbands' ties are on straight. Of course, it's all done with a patient, indulgent twinkle over our funny little ineptitudes, our obsession with nesting, our love of spending money. "You think they can't add two plus two," Stanley tells us in *Bride*, "but when it comes to weddings, they're captains of industry." Then (he tells us in *Dividend*) just when the enormous wedding, house, and decorating bills they've accumulated for you are all paid, "that's the moment they choose to let you have it" by bringing a baby into the picture, demanding more responsibility, more money from you. Pregnant women themselves have little to do; as Kay's physician reassures Stanley, "doctors and scientists have been working for hundreds of years so that women and their babies can be safe." Kay need only knit, wait, and drink her eight glasses of water a day (which she forgets to do, as husband Buckley informs us).

Kay actually grows more incompetent—and more baby-voiced—as a married, pregnant woman than she was as a bride-to-be. In the first film, she has some spunk and intelligence (although when asked what her young fiancé does she says she really doesn't know; "he makes something," she offers). In the sequel, she can't even tie her own shoes or remember to take her iron pills. Her husband "gives her money, but she keeps losing it." She spends most of the film weeping. Mother Ellie is a dour, bossy scold who rings a little bell for her maid and whose main ambitions in life are to pick out a wedding dress with just the perfect "touch of eggshell pink" (*Bride*) and "get her hooks into the baby" before the in-laws do (*Dividend*). Betty Friedan, help us!!! We need you, and *now*.

Our culture is only just beginning to pay attention to stereotypes of men and what they say about masculinity. The fact is that every female stereotype usually has a male stereotype in tow. In highly gendered worlds, "femininity" and "masculinity" are joined like the head cheerleader and the captain of the football team—opposites that belong together, couldn't exist without each other. The conniving temptress couldn't get very far in her schemes without the clueless, hormone-driven jerk to take advantage of. The wifey in the NyQuil commercials jumps out of bed in the middle of the night to bring a sleeping pill to a helpless, whining baby of a husband who, it is suggested, couldn't tell a sleeping pill from a suppository. The perfect homemaker is held up in traffic, and chaos descends on the kitchen as the bumbling dad tries to cook dinner for the family. A lot of these demeaning portrayals of men have some species of male incompetence at their core.

Father's Little Dividend is as insulting to men as it is to women. This may not be apparent at first, for Tracy's "pop" is wise, kindly, and understanding, and the family problem solver like Jim Anderson of *Father Knows Best*. *He's* the nurturer, not Ellie. Whenever his little kitten cries, he's the one who goes off to comfort her with some milk and fatherly understanding. *He* tells her to put on a winter coat when the weather is raw. While the women are clutching and clucking over what Kay should wear, where Kay should live, what the baby should be named, *he's* the one who respects Kay's wishes.

But Stanley is also—let's call a spade a spade—thoroughly pussy-

whipped by his wife. I was amazed to see how much the movie seems
to approve of this, presents it as an aspect of the natural relations between the sexes. Ellie is constantly ordering Stanley about, criticizing
his behavior. "Why must you shout? Everyone can hear!" "Why do
you have to get into such a lather that you need a drink?" "You're a
big help, I must say! Why must you leave everything on my shoulders?" Stanley just takes it, and the film thinks none the less of him for
it. He's not without his embarrassing faults, of course (it's a comedy,
after all). But allowing himself to be bossed and bullied by his wife is
not presented as one of them. In fact, it's depicted as an aspect of his
"wisdom" that he knows when it's best to simply let the women have
their way. That appears to be most of the time, especially when it
comes to matters of house, home, buying, spending, planning vaca-
tions, and so on.

Ultimately, Stanley subordinates his desires for adventure and excitement—and his more moderate material ambitions—to the values of
the nicely appointed, well-tended suburban world the women (in the
fantasy of the film, that is) are trying to construct, with frilly curtains,
up-to-date appliances, and a gurgling baby being strolled down shady
lanes. It's a world that does not come "naturally" to him as a man. In
fact, he's incompetent in it (he loses the baby the first time he takes it
for a walk, and can't even remember the color of its eyes when the
cops ask him). But it's the world he must ultimately agree to finance
and to relinquish his male independence to. There are a lot of jokes in
the movies about this capitulation, as Stanley realizes that the wedding
is going to happen exactly as his wife wants it to happen, at whatever
cost. But "girls will be girls," after all. This is the way it should be—as
Stanley learns. At the end of each movie, he's happy, having achieved
an ever deeper reconciliation with the requirements, and the pleasures,
of what is essentially a woman's world, a woman's vision of happiness.

The pattern continues through the sex comedies of the fifties and
early sixties: *Pillow Talk, Move Over, Darling, Lover Come Back,
That Touch of Mink, The Seven-Year Itch, The Thrill of It All, How to
Murder Your Wife, Who's Been Sleeping in My Bed?*, and so on. The
gender battleground depicted in these "racier" films, unlike the *Father*
movies, is not parenthood but sex and marriage. Can playboy Rock
Hudson get virgin Doris Day into bed before she gets the ring on his
finger? Can milquetoast Jack Lemmon or Walter Matthau or Tom

Ewell escape the tyranny of the married state long enough to cheat on their wives? The final answer to both questions is always: no. The woman's agenda prevails. Not that domestic life isn't mocked and lambasted, sometimes with a bitterness that belies the "good nature" of the comedy. There are lots of sardonic male speeches about the emasculation that awaits the man who falls into a woman's clutches. Here's Rock Hudson on marriage in *Pillow Talk*:

"Before a man gets married, he's like a tree in the forest. He stands there, independent, an entity unto himself. And then he's chopped down. His branches are cut off—he's stripped of his bark—and he's thrown into the river with the rest of the logs. Then this tree is taken to the mill—and when it comes out it's no longer a tree. It's the vanity table, the breakfast nook, the baby crib, and the newspaper that lines the garbage can."

But while the hero may balk and sputter and scheme to avoid capitulation, it's ultimately where his salvation lies, as is the case for Stanley Banks. All of these comedies provide the hero with a confirmed-bachelor buddy—almost always played by Tony Randall or Gig Young—whose smooth facade belies neurotic incompetence and whose seeming independence masks childlike dependence. Often, these guys have unresolved oedipal issues. Which would you men rather be—tied to the apron strings of a pretty blonde wife or cowering behind the mink coat of a domineering mommy? Even with the apron strings, these movies were on the side of the wife and against the mother. The buddy functioned as a warning against taking *that* road.

These movies are Hollywood concoctions, of course, not documentary footage on the American family. Only a minority of American families has ever had the financial security and comfortable (often luxurious) surroundings that are to this day a standard backdrop for the Hollywood comedy. *Father of the Bride* is hardly a reflection of the emotional realities of the American family; the configuration of my own family—bullying father, depressed mother, screwed-up daughters who never brought home any even remotely marriageable dates—was probably more common, even in the early fifties. Few thirty-five-year-old women kept their virginity "as sacred and well-guarded as the Pietà," as Susan Douglas drolly remarks about the Doris Day films. And looking like a masculine tree, as Rock Hudson ultimately taught us, does not necessarily mean that one is bound for the heterosexual marriage mill. (Ironically, in the sex comedies, Rock sometimes masqueraded as gay—or impotent—in order to get Doris into his bed.)

But while these movies may not tell us a whole lot about how the American family was *actually* living in the early fifties, that didn't stop their simulated realities from setting the standard for a generation growing up on their myths and images. The perfect ponytails, the crisp aprons, the heartwarming endings to trivial misunderstandings—when they didn't square with our own reality, we didn't suspect them of fakery, but assumed they existed somewhere else, for some lucky people other than ourselves. My own father was nothing like Stanley Banks or Jim Anderson. At times I was very, very glad of that. He rarely "knew best" about how to handle family conflicts and moral dilemmas (more often, he was the cause of them) but it sure was fun to go to the movies with him. Other times, I envied the safety and security of Betty Anderson's life, and I was certain there were some children who *had* dads like hers. I bet my father—who saw his own "life as a man" as a dismal failure—did too. Even Robert Young, whose life was actually nothing like Jim Anderson's—the actor suffered from alcoholism and depression and had attempted suicide—felt guilty about the disparity between himself and the image he created. But he was also angry that people took that image for reality—a confusion that continued even in Young's obituaries, when he died recently. "The perfect father figure," one described him, "always patient, understanding and sage."

It's not only the lovable figures, of course, that get mistaken for reality. So too do the bumblers, the harridans, the neurotics, the fishwives, the sissies, the ethnic and racial caricatures, the demons and devils of the myths. Some film scholars, for example, have had difficulty discerning what parts of the jailer wife of the film comedy are reflections of reality and what parts are a collective male nightmare fantasy of entrapment and emasculation. Alexander Walker, for one, describes the American domestic comedy as reflecting the "open matriarchy that prevails in the United States." (I think that even Rush Limbaugh will agree that this is a bit of an overstatement.) With this sort of confusion in mind, it's hard to sort out exactly what's going on when journalists and academics begin to complain, in the mid-fifties, about the feminization of the American male. Here's a sample, from a typical magazine piece: "We eat lettuce sandwiches and marshmallow-whip goodies concocted out of the lethal columns of women's magazines. We play Ping-Pong, we knit and crochet . . . we even (God help us) launder diapers." Is it really true that most husbands in the mid-fifties did diapers? (My own dad certainly didn't.) Or were Stanley and Ellie Banks and their ilk standing in this journalist's line of vision?

Determining how much of the "feminized male" is social actuality and how much is mythology is a question for historians and sociologists. For me, as a philosopher of culture, the "reality" of the feminized male (or the "perfect father" or the "bossy wife" or the "pert teenager") is the degree to which people believed they existed and responded to that belief. Men were soaking these images in at the same rate as women. Ellen Goodman comments that the disparity between Robert Young the man and Jim Anderson the fictional creation ought to remind people that "it isn't just women who go about our daily lives with Harriet Nelson looking over our shoulders . . . If women watching these shows wondered how these high-heeled, apron-clad mothers kept their houses so clean, did men wonder how these fathers solved all the family problems? If the girls fed on these images still carry them, so do the boys." And if girls fed on such images could become aware of their oppressiveness—an awareness inaugurated with Betty Friedan's 1962 *Feminine Mystique*—so too could men.

It's standard cultural history to mark Friedan's book as firing the first shots in what's thought of as an exclusively feminist war against traditional gender roles. But actually, it was men who fired those first shots, in protest against *their* roles. Just a couple of years after *Father's Little Dividend* had pushed the ideal of the domesticated good provider to its apotheosis, it was already being blasted apart, and a spirit of male rebellion was being stirred up. Books like Robert Lindner's *Must You Conform?* and William Whyte's *The Organization Man* warned of the dangers of male socialization and "conformity" to middle-class life. A new breed of "rebel male" actors began to define a sexy male ideal—exemplified in Brando, Dean, Clift, and the many actors they inspired—that had nothing to do with being good husband material. And perhaps the most mainstream (and long-enduring) version of male rebellion against domesticity made its appearance in 1953—just two short years after Stanley Banks happily pushed that baby carriage down Main Street—with the first issue of *Playboy*.

Hugh Hefner announced its separatist agenda in his first editorial: "We want to make clear from the very start, we aren't a 'family magazine.' If you're somebody's sister, wife or mother-in-law and picked us up by mistake, please pass us along to the man in your life and get back to your *Ladies' Home Companion*." The magazine was bitterly scornful of the kind of resolution that *Father's Little Dividend* had presented as a happy ending. "All woman wants is security," a journalist wrote in the premiere issue, "and she's perfectly willing to crush man's adventurous, freedom-loving spirit to get it."

Today, we may think of "The Playboy Philosophy," as Hefner called it, as a laughable relic of a time when you had to come up with an argument to justify naked women in a mainstream magazine. But Hefner really did conceive of himself as declaring a war of independence. He called on men to declare their "membership in a fraternity of male rebels," and described the magazine centerfold as a "symbol of disobedience." Disobedience for men, of course. The female ideal he sought to enshrine was far from a rebel. In a 1967 *Look* issue devoted to "The American Man," he told interviewer Oriana Fallaci that *Playboy* looked for "young, healthy, simple girls" who "belong to good, respectable families from every point of view. Financial, social. No, madam, we never choose poor girls. Poverty brings sadness with it, a sort of dirtiness that becomes evident even on a naked body. And the *Playboy* girls have a very high morality. After all, if the Bunnies accept a date, they lose their job. Private detectives find out if they accept a date."

Playboy, he went on, was "not interested in the mysterious, difficult women." He confessed that he didn't feel "comfortable" with "the *femmes fatales*, the old ones, or the intelligent ones," and dismissed Jeanne Moreau, Melina Mercouri, Greta Garbo, Marlene Dietrich, and Elizabeth Taylor with exclamations like "Absolutely no!" "For heavens sake!" "Don't even speak about her!" "Zero plus zero!" and so on. (The only actresses of the time that he regarded as acceptable "as women" were Ursula Andress and Brigitte Bardot "when she was young." I suspect he'd find more actresses up to his standards today in the implanted and liposuctioned nineties.)

Playboy's style of male rebellion was better suited to most American men than Kerouac's beatnik or Mailer's "hipster," who achieved their manhood by living a life of danger and marginality, even madness.

Playboy offered an escape from domesticity that did not require giving up bourgeois comfort. Far from it. Imported liquor, great stereo equipment, mile-long cars, and snazzy bachelor pads were the accouterments of the Playboy lifestyle. The products sold, and the antidomestic "philosophy" took. By 1955 and *Rebel Without a Cause*, Jim Backus's ineptitude as a father and man was signaled to the audience by the fact that he wears an apron and performs domestic duties around the house. When he drops the tray of dinner he's bringing to his wife's room and scrambles to clean it up before she can see it, his son Jim (James Dean) is embarrassed by his father's lack of manliness. Spencer Tracy, subjected to the same kind of feminine rule in *Father's Little Dividend*, is doing what men are supposed to do: grumble, sigh, and give in. He's none the less "masculine" for it.

The recognition that the birth of contemporary gender protest includes both a male rebellion and a female rebellion certainly makes for strange new bedfellows. Suddenly, it seems, the mid-fifties emergence of *Playboy* on the one hand and the appearance of Betty Friedan's *Feminine Mystique* in 1962 on the other no longer can be seen as polar events. The masculine ethos that Hefner encouraged, no less than Friedan's call to feminist consciousness, was an "I'm mad as hell and I'm just not going to take it anymore" response to the happy homemaker/good provider mystique that the early fifties is famous for. Yes, Hefner was astoundingly, gaggingly sexist in his ideas about women. And it's also true that in the sociological tracts and movies advocating male rebellion from domesticity, women were frequently blamed for all of men's woes. But then, so too did early second-wave feminists often blame men. The development of a systemic, institutional, *cultural* understanding of gender has been a long, hard time coming—and we still haven't nailed it. It's so much easier to blame than to analyze. Analysis is hard, it's complicated, and it disturbs the comfortable simplicity of familiar worldviews.

The Stanley Banks version of masculinity began to self-destruct almost as soon as it was articulated. And we've never really gone back to it fully, though *The Cosby Show*—which was able to freshen up the bossy wife with the ethnic shadings of sass and sex appeal that Phylicia Rashad brought to her role as Clair Huxtable—came close. The 1991 remake of *Father of the Bride*, starring Steve Martin and Diane

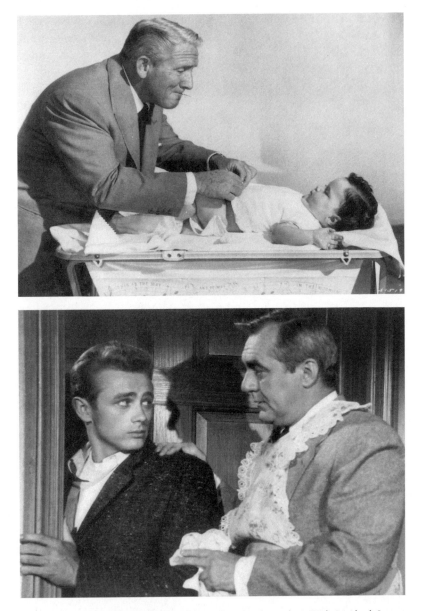

Changing views of domestic manhood: in a publicity shot for *Father's Little Dividend*, Spencer Tracy happily changes diapers; in *Rebel without a Cause*, Jim Backus's frilly apron tells us that being a helpful househubby is unmanly. (*Photofest*)

Keaton, updates the 1951 version in a variety of predictable ways (daughter now has master's degree, plays basketball with Dad, gets upset when husband gives her a blender). But what's most striking is the update that *isn't* so "politically correct": Diane Keaton is softer, sweeter, more respectful of her husband, and ultimately much more femininely manipulative than Joan Bennett was.

The male rebellion of the fifties was not *feminist*, by any means. But revisiting that rebellion with *Father of the Bride* in mind made me remember *why* I had once felt Norman Mailer to be such an ally and inspiration for my own rebellion. Stanley Banks and Ellie Banks were a team, playing a game whose rules I found as difficult to conform to as Norman Mailer and Philip Roth did. When these writers (and others) blasted away, even in misogynist fashion, against the false mythology of fifties family happiness, they were giving passionate voice to much that I felt but that women did not yet have their own accurate, nonmisogynist language—or permission—to express. Lots of other women of my generation, as Wini Breines has chronicled, felt the same way. "I wanted wildness, originality, genius, rapture, hope," recalls writer Annie Dillard, "I wanted strength, not tea parties." At that time, writers like Mailer were the ones *not* serving tea.

The fact is, too, that it was not just men who found the dutiful, obedient husband an unappealing masculine ideal. In the same *Look* issue in which Hefner was interviewed, Patricia Coffin sends this "message" to "The American Man":

> You are too nice. You have been too nice for too long. Some of you are decent-nice. Some of you are nasty-nice. Others among you are mealy-mouth-nice, conformist-lazy and some, heroically-nice. But niceness does not become you. As providers, you American men are unique. As doers, no one can beat you. As lovers, you are gauche but great. Yet you take the cake when it comes to being pushed around by your women. Do you want a world of she-men and he-women?

Today, I am appalled by these words, which Coffin wrote in 1967, at the very brink of the second wave of American feminist activism. Even by 1969, I would already have found her sentiments unacceptable in many ways. But the fact is that I rarely went for "nice" men. And believe it or not, I learned not to from the movies of the fifties.

if it's punishable, it must be sexy

Margaret Mead declared, at the dawn of the sixties, that we had overnight "jumped from Puritanism to lust." That was (and is) the story as commonly told about the "sexual revolution" of the sixties and seventies, and looking at movies it certainly seems to be an accurate assessment. The fifties were known as the "Plague Years" of Hollywood self-censorship. It's hard to believe it in 1998, but Otto Preminger's 1953 *The Moon Is Blue* was rejected by the Hollywood Production Code for using the word "virgin." In 1956, Preminger's *Man with the Golden Arm* got nixed for its realistic depiction of drug addiction. "Indecent or undue exposure" of the body was prohibited. Official and unofficial lists of dirty words and dirty deeds made distorting revisions necessary when Broadway plays, even classic literature, were transposed to film. Brothels could not be identified as such, open-mouthed kisses were prohibited. The word "abortion" could not be spoken (although of course the horror of it could be alluded to, just as comedies could be constructed around the heroine's dedication to preserve that other unmentionable until marriage).

By the mid-sixties, the Code had been eroded by the "foreign invasion" of sexually franker European films, and by the willingness—actually, eagerness—of audiences to see U.S. movies that had been denied a seal of approval. The movie industry knew too that it had better do something to compete with television (which still insisted on married couples sleeping in separate beds). In 1965, *The Pawnbroker*, which included a scene in which a prostitute (Thelma Oliver) exposes her naked breasts, was approved on grounds of artistic necessity and cracked the body barrier. "Film Art Requires No Bra," *Variety* headlined. Probably, the fact that Thelma Oliver was black—and thus less likely to stir the protective responses of the Breen office's all-white officers—helped things along. It's clear, though, that that bra was going to come off soon, and it wouldn't be long before the "art" part was no longer required either. The next year, Mike Nichols's screen adaptation of Edward Albee's *Who's Afraid of Virginia Woolf?* did the same for language—and prompted a major revision of the Code itself. In 1966, *Variety* magazine declared: "The Code is Dead."

But actually, the history of Hollywood's regulation of sexual material is more complicated than a story of stifling self-censorship relieved by the fresh winds (and commercial avidity) of the sixties. The story begins in 1915, when the Supreme Court ruled that movies were not covered by the First Amendment and in response states began to set their own censorship laws. The language of those laws was often loose, their emphasis not on specific behavior or words, but on the potential effects of films on viewers. New York's 1921 law, for example, ruled against licensing any film that "would tend to corrupt morals or incite crime." This kind of language immediately suggested a clever strategy to filmmakers like Cecil B. DeMille. If you showed sin, but made sure to have God punish it, how could you be accused of corrupting morals or inciting crime? Indeed, it could be argued that by making sinners roast in hell (or pay for their crimes in other hideous ways), you were discouraging immorality, performing a public service!

Biblical epics were naturals for this sort of thing. Their orgies and other sexual excesses—as well as their grisly punishments—were taken from the most morally unimpeachable source. One could be sexually titillating and moralistic at the same time, and all in the name of the enduring lessons of the Good Book. How could anyone object? Ultimately, many moralists did, arguing that the biblical settings were just a pretext for the depiction of depravity (the punishments were often as titillating as the sins!), and in 1922, Presbyterian elder and former postmaster general Will Hays was drafted to head the newly created Motion Picture Producers and Distributors Association of America. Initially, the organization's role was mainly public relations, designed to head the external censors off at the pass. But gradually it evolved under Hays's leadership into an organ of rigid self-regulation (renamed the Production Code Administration in 1934, and given a new moral watchman, Joseph Breen, a former Roman Catholic newspaperman), whose Code grew ever more specific in legislating what could and couldn't be portrayed.

But even in the heyday of the Code, that old biblical-epic strategy— make the sinner roast in hell for the moralists, but be sure to show him sinning first for the sensualists and realists—was often employed. In Breen's censorship manual, it was permitted by the "moral compensation" clause, which improved on the old informal rules by insisting only that the punishment be equal to the sin. The punishments may not have carried quite the punch that Breen intended, however. The sexuality of the fifties, like the Victorian era, was much more complicated than Margaret Mead's "Puritanism" versus "lust" dichotomy suggests. Yes, it's true that during much of the fifties, sex (especially sex for sex's sake, not aimed at marriage and family) was portrayed as nasty, dirty, evil. But, as was true for the Victorian era too, moralizing against illicit behavior and eroticizing it were just a stone's throw apart. It's almost impossible to portray something as forbidden without also portraying it as enticing. The sex might be stuffed behind some socks in dad's top drawer—that's where my father kept his copy of the famous calendar that Marilyn Monroe posed naked for (unfortunately for my bank account, it's since disappeared)—but that only made it all the more exciting for me to sneak into that drawer, as I did time and again, hypnotized by the cherry color of her nipples and her smooth, indrawn stomach.

Movies of the fifties—even many "family" movies like *The Ten Commandments* (1956)—are far from sexless. In fact, I was shocked, when I rented the film recently, to discover just how sordid it was. I originally saw the film when I was nine, at a drive-in movie with my entire family. We had gone to a drive-in in deference to my Orthodox Jewish grandmother, who wouldn't go to movie theaters. We didn't think she should miss this particular movie, though. Even if she couldn't understand the dialogue (she spoke only Yiddish) we knew the visuals—the parting of the Red Sea, the burning bush—would speak to her, so we all (father, mother, grandmother, my two sisters, and I) piled into the car. A real family outing. And for a real family movie. That's how I remembered the event.

I *didn't* remember the following speech, put into the mouth of pharaoh-to-be Ramses (Yul Brynner) and spoken to Egyptian princess Nefretiri (Anne Baxter, in the days before white directors learned that Egyptians were "people of color"). Nefretiri is in love with Moses, but Ramses is determined to have her as his wife. Holding her arm, looking into her eyes with cold, undisguised lust (something Brynner did exceptionally well), he informs her of his intentions:

"I will have all of you. You will be my wife. You will come to me whenever I call you, and I will enjoy it very much. Whether you enjoy it or not is your own affair. But I think you will . . ."

Watching the movie again, I wondered what scenes like this might have communicated to my prepubescent (but fast becoming boy-crazy) self. I knew that Charlton Heston as Moses didn't do anything for me at all (I was surprised to see how nice his body was in the film, because I remember him as repellent to me). Chuck Heston was boring. Too pompous. Too stiff. Too wholesome to be sexy. Too wholesome to be sexy? Where did I get that idea? I realized I got it from the movie itself. It gave sexual atmospherics, knowledge, consciousness—and promise of pleasuring the woman—to the sleazy Ramses. Moses? He was the movie's hero. In mid-fifties movies, pursuing sex, caring about it more than saving the world, was not considered an activity appropriate to such a noble being.

Heroes *could* be undressed, racked, whipped, and stripped of their dignity, however, and in a highly eroticized way. And they were, often. (Except for the occasional punishment of a slave girl, it's men, men, men whipping and being whipped by each other.) Thanks to fifties biblical epics, my ideas about sex were tinged with S&M from the start. Those movies are full of captive, semi-naked, spread-eagled heroes. Their masters stand there, also bare-chested, legs apart, stomachs sucked in, stiff and as pumped up as they could make themselves in those days, wielding their instruments of torture. *The Ten Commandments*, in particular, is chockablock with bondage paraphernalia— wrist cuffs, arm cuffs, slave bracelets—and beatings are never just beatings, they are always occasions for sadism. (*The Ten Commandments* also features one of the most hilariously symbolic male contests, involving who—Moses or Ramses—can successfully erect an enormous, phallic obelisk. Ramses lost the contest but went on to have a condom named after him.)

A film whose S&M ambiance made a big impression on me—I was thirteen—was Stanley Kubrick's *Spartacus*. In the movie, Kirk Douglas's gleaming body is on full display, as slave dealers and pampered Roman royalty inspect his suitability as a potential gladiator. The first scene finds him stretched on a rock, nearly dead; he's spared only because slave dealer Peter Ustinov, inspecting Spartacus's teeth and other parts, sees that he's special. So does rich Roman bitch Nina Foch, who chooses him to fight to the death with a statuesque black gladiator (Woody Strode). "Tell them to take off those suffocating tunics," she instructs Ustinov.

The other male "sex object" of the movie—Tony Curtis's Antonius, the "body servant" of Crassus (Laurence Olivier)—flees his master when he realizes that Crassus expects more than "the singing of

Biblical S&M: Kirk Douglas enslaved and on display in *Spartacus* (*Photofest*)

songs" from him. In a notorious scene—cut from the version released in 1959 and restored in the current video release—Crassus, who is being bathed by Antonius, asks Antonius for his robe and tells him that he is one of those people whose taste "includes both oysters and snails." (It's after this that Antonius, not wanting to be eaten raw on the half shell, runs off and winds up in Spartacus's band of rebels.) It wasn't the first homoerotic bondage moment to come from fifties Hollywood. In *The Ten Commandments*, bare-chested Joshua (John Derek) is tied to four pillars and beaten by a sadistic Boris Karloff, who remarks that it "is a pity to kill so strong a stonecutter."

Stanley Kubrick's *Spartacus* appeared in 1960, and in many ways is a transition movie, both sexually and in terms of cultural politics. Kubrick (and Dalton Trumbo, in his first post-blacklist screenplay) use *Spartacus*'s moments of sexual degradation and bondage for high instructional purposes. Spartacus learns, from the black gladiator who refuses to kill him, that dignity and brotherhood are worth more than life. His experience as a slave also makes a protofeminist out of him.

When Ustinov brings slave girl Varinia (Jean Simmons) to service Spartacus, he realizes (after a bit of instruction from her) that he and she are in the same degrading position, and refuses to have intercourse with her so long as she's his sex slave. The "politics" of *Spartacus*, filmed just as one decade gave way to another, and by a director who was to represent the next wave in so many ways, were avant-garde for an American film.

When I was thirteen, the political messages were largely lost on me. But I was enthralled when Spartacus and Varinia, who get together again on the road, joyously engage in what in the fifties was "premarital sex." And she even has his baby—"out of wedlock"! These behaviors, which were to become mundane movie fare in just a few years, still pulsed in 1960 with transgression—and thus the suggestion that the way to have more fun (Varinia and Spartacus seem as sexually satisfied as a couple could possibly be) was to be an outlaw to convention. The boys with whom I was friends got the same message; even as young teenagers, they began to be scornful of the girls who cared about proms and pleasing teachers, fantasizing about a wild rebel girl—preferably with breasts as big as Elly May's in *The Beverly Hillbillies*—to cut classes and experiment sexually with. (At the time, I could only offer the disdain for proms and willingness to skip school.)

Looking forward to the sixties, the sex between Spartacus and Varinia was coded as playful, healthy: frisky skinny-dipping, bundling under big blankets. Much more common accompaniments to sex in fifties movies were visual and other signals (sultry jazz "sex music") that the characters were moving out of the sunlight and into a dark, dangerous underworld. Death, jail, insanity, or some other punishment was likely to follow ecstasy, and not just for those deemed "evil" by the censors. Even those who descended into sex innocently and in love with each other usually got punished—like Tup-Tim (Rita Moreno) and her lover in *The King and I* (1956), whose forbidden "kiss in the shadows" is paid for by his death and her recapture. Did this ending function as a deterrent to my desire to find someone to meet in the shadows? Hardly. What it did do was to add a dimension of danger to my developing definition of sexual excitement. "Definition," of course, is too cerebral a word; these equations were learned in my body. For me, the message of grim endings to forbidden sex was not "don't have sex or you'll get punished" but "if it feels punishable it must be sexy."

I'll never forget Yul Brynner's body when he's about to whip Tup-Tim. He flings off his shirt, plants his legs about a mile from each other, and raises the whip in a moment of pure, unadulterated bodily exhibition. Whose artistic choice this posture was—the director's or Brynner's—I don't know. I do know that it's impossible for the viewer to avoid the visual association between Brynner's strong, sexual beauty and the dynamics of dominance and submission the scene embodies. Let me put this more concretely. Although the king, who is not really a Simon Legree, is unable to use the whip on Tup-Tim, the fact that he's got it in his hand and has the power to use it is . . . sexy. It didn't have to be. The king didn't have to go all Promethean and splendid as he raised that whip. They *made* it sexy.

I had a terrific crush on Yul Brynner. Yes, even before Edd ("Kookie") Byrnes there had been Yul. My attachment to Yul was different from my later (more "mature"!) crushes on teenage media figures like Kookie and Fabian. With them, my sexuality was already on its way to becoming "normalized"—I fantasized about being a girlfriend, thought about going on dates with them, and so on. Yul belongs to an earlier period of imprinting, when dreams of domestic bliss with a boy had not yet infiltrated my id. His bare-chested moments in *The King and I* and *The Ten Commandments* (not to mention his body-clinging black Russian garb in *Anastasia*, during which he again carries some kind of whip, or riding crop) were a fetish—not entirely clear or understood by me, of course—of my developing sexuality. His eyes were piercing, his lips were full, his bald head and slightly Spock-like ears were menacing. He thrilled me. I even stole a 45 rpm boxed set of the soundtrack to *The King and I* (stuffed it inside my winter coat and ran out of the record store) partly for the music but also in order to look at his body on the glossy cover of the album.

I stole Harry Belafonte's *Calypso* album too, and adored watching him on Ed Sullivan, his shirt open to the waist. I didn't realize at the time that the bodies of black men and other "exotic" types like Brynner—who was Russian—were often presented in a more explicitly sexual way than permitted the white guys. Even Sidney Poitier was first a slinky, T-shirted hoodlum (in *Blackboard Jungle*) before he became Mr. African-American Dignity in later movies. Ethnic exoticizing and caricaturing aside, some of these actors *did* have a more seductive way with their bodies, voices, and movements. Racial differences, as I'll explore more in another chapter, do matter a great deal in men's attitudes toward the presentation and "styling" of the body. Brynner brought an unusual degree of erotic consciousness to his parts, a sense of body awareness that few men in the movies exhibited freely in those days. I wasn't surprised to find, in a recent book, *Pioneering Male Nudes*, that Yul had posed for George Platt Lynes. The 1942 photograph included in the book shows him reclining against a wooden platform, utterly "nonchalant"—as the book's author comments—in his naked beauty.

reel boys and real boys

Yul Brynner was not date material, but he definitely was sex material. That distinction, generally thought to be an exclusive feature of male-think (girls you marry and girls you screw around with), was—perhaps inadvertently—taught to girls too in the fifties, with significant consequences for young men. With the world now becoming divided into "nice" boys and exciting boys, qualities that a decade before were assumed to appeal to women—courtesy, kindness, stability, integrity— began to give off the stale odor of the square and boring. Of course, the movies had always had their share of thrilling, dangerous men. But it wasn't until the early sixties that the style of the cultural and sexual outlaw began to become a model for middle-class teenage males.

Yul Brynner was not the paradigm of this new style, however. The boys I was friends with in high school—all straight (or so they all thought) at the time and, except for one close black friend, white— didn't emulate Yul; he was too exotic, too explicitly erotic for them. And, of course, he had no "message." Actually, my friends didn't emulate movie-star images at all, but jazz musicians like Charlie Parker and Thelonious Monk, literary mavericks like James Joyce, whose *Portrait of the Artist as a Young Man* turned their adolescent identity crises into a new kind of masculine poetry, and existentialist machomen like Norman Mailer, who celebrated the "psychic outlaw" who resists "slow death by conformity" by taking "an uncharted journey into the rebellious imperatives of the self." The disaffected heroes of Jack Kerouac's *On The Road* and Joseph Heller's *Catch-22*. Lenny Bruce's caustic, anti-establishment humor. (My boyfriends were thrilled when he got busted in 1962 for saying "motherfucker" onstage.) Allen Ginsberg's "Howl." Norman O. Brown's *Life Against Death*. These early countercultural statements, calling for a robust collaboration of political critique and sexual liberation, were defining a new kind of rebel virility for the boys that I knew. The civil-rights movement, campus activism, and anti-war politics would soon enlist that virility in social action.

My boyfriends in high school would have been utterly scornful of the notion that movie stars had any impact on their physical or psychological self-image. And in terms of direct influence, they would have been right. Among popular entertainers, the only one whose style of masculinity they consciously emulated was that of folk singer Bob Dylan, who combined poetry, domestic rebellion, and anti-establishment passion with lousy posture, a Salvation Army wardrobe, a voice that defied convention, and a sense of indifference to his audience, even when bellowing lyrics of protest. My friends slouched like Dylan, they mumbled like Dylan, they scoured the Newark thrift stores in search of Dylan-like clothing. What most of them didn't know, however, was that young Robert Zimmerman (Dylan), and many other singers and public politicos of the sixties, just slightly older than the boys I knew, had gotten *their* inspiration from two movie stars of the fifties: Marlon Brando and James Dean.

Dean had the more obvious and dramatic impact, not only because of his shocking death (he was killed instantly in a car crash just three days before the opening of *Rebel without a Cause*), but because his style was explicitly that of alienated *youth*. Graham McCann comments: "[Dean's] image appealed romantically to the 'hurt' children of post-war upheaval. His callow face had the flickering mobile uncertainties of adolescence continually passing over it; his speech was a secret mumble; his myopia gave him a naturally withdrawn, wincing look; his dress was self-consciously sloppy, purposefully unkempt. He seemed, more so than [Montgomery] Clift or Brando, to merge with his screen image of the hyperactive, androgynous, impatient teenager."

Dean died and *Rebel* opened in 1955. Almost immediately, fan clubs sprang up all over the country, as the date "9/30/55" was carved into school desks and printed on T-shirts. The boys around my age—the first wave of the baby boom, born in 1946 and 1947—were only eight or nine years old, more likely to identify with the cartoon perils of the Road Runner than with the teenage *Angst* that Dean projected. But their future hero Bob Zimmerman was already a young teenager, and "the moment [Zimmerman] saw *Rebel without a Cause*," McCann reports, "he adopted James Dean's blue jeans and boots, his slouch and his smirk." He completed his outfit with a black leather jacket in imitation of Marlon Brando in *The Wild One*.

It is Dean, not Brando, who is thought of as the more influential figure among baby-boomers and their pop idols, and not without some justice. (Elvis Presley, it is said, watched *Rebel* a dozen times, and knew its dialogue by heart.) But actually it was Brando who originated the new elements of masculine style with which Dean became associated, and which Dean consciously tried to copy—from Brando's voice, walk, gestures, and clothing to his iconoclastic opinions. Dean wasn't alone. According to fellow actor Anthony Quinn, once other young actors had seen Brando as Stanley Kowalski in Tennessee Williams's *Streetcar Named Desire*, they all started "behaving like Brando." He "fucked them all," Quinn reports. "The whole thing up until then, everything was proper. Robert Taylor, Tyrone Power, Van Johnson, and along comes Brando. It was the character, his Napoleonic Code speech—'A man is a king, by the fucking Constitution he's a king!' That statement turned the whole world around."

Actress Sally Quinn, who dated both Brando and Bob Dylan in the mid-sixties, says that Dylan "adored, idolized ... was terribly attracted to Marlon." So was Black Panther leader Bobby Seale, although without the sexual interest that Dylan apparently felt (according to those, like Quinn, who knew Dylan intimately). Seale recalls that he "had always admired Marlon from the time I was sixteen and saw *The Wild One*" and identified with his "rebellious streak," stories about which he would avidly gobble up in the gossip magazines. James Dean's own hero-worship of Brando bordered on (perhaps crossed over into) obsession, as he followed Brando down the streets of New York City, made late-night telephone calls, even showed up at a party dressed in an exact replica of Brando's *Wild One* outfit (Brando himself thought this "sick").

It was thus Brando, not Dean, who was the first to translate the emerging ideology of the male rebel into both screen imagery and public personae—and long before he put on that leather jacket in *The Wild One*. Even before he became a star, Brando flouted the establishment with his body, appearing at the Russian Tea Room in visibly ripped shirts, attending New York parties wearing sneakers and an old raincoat, standing alone in the corner all night. This kind of behavior led the press to tag him (much to Brando's chagrin) as uncivilized, a primitivity they loved to illustrate and vivify with descriptions of him scratching himself on the butt, picking his nose, walking barefoot into restaurants. "Rude, moody, sloppy and prodigal," is how a 1950 *Life* magazine article describes him. ". . . A harlequin who had not been house-broken. Not in recent memory had the town [Hollywood] been confronted with an actor who treated almost any rudimentary standard of social behavior with such utter and childlike disdain."

The press, as we'll see, was not entirely fair to Brando. But accurate or not, a certain imagery had taken root in the culture. Brando passed it to Dean, Dean passed it to Bobby Zimmerman, and Bobby Zimmerman (as Bob Dylan) passed it to the boy I was in love with in high school. "I'm fantasizing about him in his room," I wrote in my diary, "in his undershirt and dungarees. His hair is long as hell and he is beginning to get that derelict look he so often gets when he is kind of overgrown . . . His room is full of all sorts of shit—old bottles of VO, scratched Bill Monroe, Blind Willie McTell, Monk, Evans and Charlie Parker disks, a few baroque and madrigal type stuff for class, books strewn around, a huge 'Soft Shoulders' sign, a bus stop from Lyons and Parkview, and a battered maple dresser with a host of Trojan-Enz scattered across the top . . ."

Ah, romance . . . ! (It was 1962.) When I wrote these lines, neither I nor the bad-boy object of my (unrequited) affections had seen *A Streetcar Named Desire*. Marlon Brando was about as far removed from my fantasy life as Charlton Heston. In those days, I would have regarded both actors as belonging to the same earlier, uncool generation. But in 1998, teaching a course on film and thinking about images of masculinity, I realized how wrong I was. My twenty-something students, too, were transfixed by their first viewing of *Streetcar*. Up until then, they had only known Brando in his Godfather and post-Godfather incarnations. They were unprepared for Brando's sculpted body, but, more significantly, for the male beauty, power, and sexuality he projected. "WOW!" they wrote in their journals. In a culture now inundated with muscular male sex-gods, that "wow" deserves further exploration.

kowalski and sons

Who can forget Stella Kowalski's (Kim Hunter) slow, sultry descent down that curved iron stairway outside the tawdry Elysian Fields tenement, in Elia Kazan's 1951 film of Tennessee Williams's *A Streetcar Named Desire*? At the bottom of the stairs is her husband, Stanley, crying for her like a baby that's lost its mother, his muscular yet graceful back exposed, his T-shirt in shreds. A few moments before, he had beaten pregnant Stella in a rage, ostensibly over her sister Blanche's (Vivien Leigh) interfering with his poker concentration with radio music, but more deeply over Blanche's growing influence on Stella. Now, realizing what he's done, he's distraught, beside himself with remorse. As Stella reaches the bottom of the stairs, he drops to his knees before her, and buries his head in her body. "Don't ever leave me, baby," he pleads; her hands run down the length of his back, then grasp his head, his hair, as she kisses him passionately, voraciously.

The idea that Stella's lust would overpower her, even after she's been pummeled by Stanley, was too much for the Catholic Legion of Decency, who threatened to give the film their "condemned" rating unless both elements were modified. In the version that premiered in 1951, a few choice cuts—made by the studio behind Kazan's back (a restored "director's cut" video was released in 1994)—mitigate the violence in the beating scene, as well as the sultriness of Stella's growing, heavy desire (at times, in the restored version, she looks downright slutty) as she listens to Stanley's wailing, leaves the room, begins to go down the stairs. The ending of the play had already been altered for the Breen office, which insisted that Stanley's rape of Blanche (which Williams refused to eliminate) would have to be punished in some explicit way. Williams and Kazan had reluctantly agreed. Thus, the movie ends with the suggestion that Stella, newborn baby in her arms, will no longer listen to Stanley's cries, will not return home despite the sexual magnetism between them.

I doubt that many female viewers easily bought the morally enhanced ending of *A Streetcar Named Desire*. It's what Stella *should* do, of course, but the film—even with the cuts—casts doubt on its plausibility by presenting such a compelling psychosexual portrait of *why* women stay with men like Stanley even after they do awful things. Brando's interpretation was key here, in giving full expression to Stanley's dependence on Stella, his almost childlike attachment to her, his desperation to make things up with her when he's gone too far. After he beats Stella, his remorse is huge and his fear that he has lost her—even for a night—intolerable. Those famous cries—"Stella! Stella!"—are not the grunts of a caveman but the screams of an abandoned infant who feels he will die if he isn't fed. ("He was as good as a lamb, and very ashamed of himself," Stella tells Blanche the next morning.) Yes, there's the sexual bliss, the "colored lights" (as he puts it) they get going together. But Brando's interpretation astutely enhances another dimension—one that contemporary experts on spousal abuse would concur with: Stella is hooked on Stanley's need for her, and that, unfortunately, is reaffirmed even more strongly after he's been violent and desperate to make amends.

However women viewers may have responded to the ending, they (and not only they) undoubtedly went away from the movie with a set of powerful images emblazoned on their sexual imaginations. That first shot of Brando, taking off his bowling jacket, revealing a wet, clinging short-sleeved T-shirt stretched over the most beautiful male chest ever seen before or after on the screen. ("Isn't he wonderful-looking?" Stella swooned to Blanche. He was, oh he was.) Stanley at the bottom of the stairs, of course. Stanley walking down the street, this time in a sleeveless T-shirt and a funny little cap, dirt all over his face, looking like a little boy who's been playing in the mud. He passes an open window, which carries Blanche and Stella's discussion to his ears. "He's common . . . an animal . . . subhuman. . . . Don't hang back with the brutes!" Blanche has urged Stella. He appears at the doorway, unfazed, dirty and sweaty, whistling. He pops the top off a beer, a grin on his face. Stella jumps into his arms. He knew she would. You could tell by that grin, and his silence, that he was confident he had the more powerful argument.

Stanley knows the effect the sight of his body—and the memories, suggestions of the pleasure it can provide—will have on Stella. And Brando, too, was well aware of the sexual heat projected by the gorgeous physique that he cultivated for the part. "He was just about the best-looking man I had ever seen," said playwright Williams. No wonder—Brando worked out hard and daily at the gym, viewing the right torso as an essential element of the character. For Stanley's clothing in the play, costumer Lucinda Ballard made two major innovations, re-

The Look that reverberated across generations: Marlon Brando as Stanley Kowalski
(*Photofest*). A contemporary descendant in
a DKNY Jeans ad evokes both Dean and
Brando

tained in the movie, whose repercussions we are still experiencing to-day. In the early fifties, the T-shirt was typically worn loose, with sleeves down to the elbows, and jeans were cut baggy. Ballard shrank the T-shirts to cling, and fitted the jeans while wet, pinning them tight around Brando's naked groin and bottom (Brando insisted on wearing no underwear). Biographer Peter Manso reports that Brando "almost went crazy" when he saw how he looked in the skintight shirt and jeans: "This is it! This is what I've always wanted!" he exulted. Not just Brando. Brando's look in *Streetcar* (shortly to be copied by James Dean, Paul Newman, and others) became the style for sexual macho in many gay male circles and a required uniform for many would-be teen rebels.

Mr. Breen, in thinking his ending had removed any suggestion that *Streetcar* condoned Stanley's primitive sexuality, hadn't yet learned that the image, not the moral, was becoming the message in Holly-wood. The second trailer for *Streetcar*, issued after the movie had won five Academy Awards and after Brando (though he lost the Oscar to Humphrey Bogart) had become the hottest new actor in town, undid everything that the Legion of Decency and the Code had wanted to ac-complish. "Fighting, lusting, loving. . . . Never for a moment less than completely alive, a man who had two women living in his house, re-acting to his savage appeal!" the voice-over exulted, while the viewer was led through a pastiche of the most violent, erotic scenes in the film, including the ones that had caused censors the most trouble: the scene where Stanley tears up the house, Stella on the stairs, the rape scene. It even included the moment just before the rape, when Stanley tells Blanche that he'll give her the "roughhouse" she wants (which, out of context, could be taken as the truth about what Blanche wants). The trailer culminates in the shot of Stella being carried upstairs in Stanley's arms. So much for muting the brute's erotic appeal to the ladies.

The narrative of the original trailer—run before the movie had been released—told a very different story. Like the play itself, its focus is not Stanley but Blanche, who "wanted so much to stay a lady," who "tried to give her heart honestly, completely . . . fighting to rise above her past," but whose "story—every searching chapter—was written by men, who taught her to trust, love, hate." The announcer then leads us on a mini-tour of the men in Blanche's life (the young man who killed himself, Mitch, who wanted to make an honest woman of her, etc.), ending with Kowalski—"the brute who lied and cheated." The chosen images were much, much tamer, and ended with a shot from the very beginning of the film, of Blanche, just arrived in New Orleans, looking lost and bewildered at the train station. Watching this trailer and comparing it with the second was fascinating. Like the second trailer, it puts a very particular, half-truth spin on the movie. But the halves are polar opposites in what they choose to emphasize and omit.

The earlier trailer presents *Streetcar* virtually as a traditional woman's movie—not just because it focuses on Blanche, but because the language is that of heart, heartache, loss, longing, and betrayal. The later trailer is all sex, brutality, "a story ripped from the fabric of life, as earthy and violent as its unforgettable star." By "unforgettable star," they now meant Brando, not Leigh (who won the Oscar and whom the original trailer had described as "the outstanding actress of our day"). In one of the scenes in the second trailer, Stanley tells Blanche and Stella to "remember what Huey Long said—every man is a king, and I'm the king around here." That's the message of the trailer too. Created after Brando's electrifying Stanley had made "more impact in a single role than any actor in history," it not only bumps Blanche from the center of the movie but displaces her values too—in favor of Stanley's. The trailer is one of the purest celebrations of macho, primitive masculinity I've seen.

Warner Brothers, in emphasizing the glory of Stanley's "savagery" while minimizing the serious themes of the play, was undoubtedly hoping for better box office revenues. Hollywood was just at the beginning of the era of blockbuster commercialism that would make the "woman's picture" and the "problem drama" aberrations rather than staples, as they had been. I believe, too, that something of the fifties male rebellion was beginning to manifest itself in the Conan the Barbarian pitch of the second trailer. This was not exactly the side of Brando, however, that made women "breathe heavy in his wake," as George Glass (who produced *The Men*) put it in a 1953 article in *The Saturday Evening Post*. Glass recalled, in Raymond Chandlerish color, how "when the femmes saw him in [the stage production of] *Streetcar*, they crawled right up the theater walls. They say there's a heavy shot of masochism in every dame. I wouldn't know. I do know that they shivered and wriggled at the way he talked, although some nights you could hardly hear him past the tenth row." Glass has the shivering and heavy breathing right, but in this dame's opinion he's wrong about the masochism. It wasn't Stanley's brutality that excited women, it was the intensity of his passion. Watching him in *Streetcar*, a female member of the Actors Studio (where Brando studied) said, was "like being in the eye of a hurricane." The rape scene is Stanley at his least sexually appealing, because he reveals how compassionless and cruel he really is. The Stanley women thrilled to was not the man who selfishly shattered other people's bodies and dreams, but the man whose sexual vitality was so unhobbled that he smashed all the lightbulbs in his house on his wedding night. There's a difference. Women responded to the Stanley who ran with the wolves, not the brutes.

So, it seems, did Brando, who was hurt when magazine articles described him in terms that were suitable for boorish Stanley but not, he insisted, for him. "What I have to say is either misinterpreted or misunderstood, and I always feel betrayed afterward," he told *The Saturday Evening Post* in 1953. "But I'm building armor against this sort of thing so they can't hurt me anymore." He voiced disgust with Kowalski's cocksure brutality. "Kowalski was always right, never afraid," he said. "He never wondered, never doubted himself. His ego was very secure. And he had the kind of brutal aggressiveness I hate. I detest the character." Of course, Kowalski, as Brando played him, was *not* secure. Aggressive and brutal, and seemingly without self-doubt, yes. But secure? The brilliance of Brando's performance was his willingness to expose the blubbering baby inside the caveman. When he cried for Stella, he scrunched his face up like an infant whose bottle has just been taken away, and let the tears flow. No man had ever cried onscreen in that way before.

Brando himself was unashamedly sensitive. "If there are 200 people in a room and one of them doesn't like me, I've got to get out," he told *Life* in 1950. A famous story, repeated over and over again in magazines, describes how copiously he wept during the scene from *The Wizard of Oz* in which Judy Garland sees a distant vision of her aunt. During the shooting of *The Men*, he reprimanded one of the producers for not loving his cat enough. Whether or not these stories were true, they too contributed to Brando's image and developed it in a very different direction from that of macho man.

Graham McCann accurately describes Brando as one of the "great promissory icons of 1950's male sexuality." That promise branched out in several contradictory directions. Brando spawned many imitators—Lee Marvin, Clint Eastwood, Steve McQueen—who capitalized on the tough-man aspect of Brando's image. But another branch would lead away from pure, action-oriented toughness and toward a greater range of emotional expression. Fellow actor Roy Scheider once said Brando could "outfeminize any woman in any scene." I'm not exactly sure what Scheider had in mind here, but I do know that even as Kowalski, Brando has nothing in common with the cold, clenched-teeth, relentlessly heterosexual Dirty Harrys of the seventies and eighties.

James Dean, most notably, took Brando's legacy in the direction of this more sensitive, vulnerable, and ultimately sexually ambiguous masculinity. "Before James Dean," remarked Sal Mineo, who played alongside Dean in *Rebel Without a Cause*, "you were either a baby or a man. There was nothing in between." Actually, as we've seen, it was Brando who first challenged that opposition. But Dean elaborated the alternative—the child-man straddling the poles of neediness and manliness—in terms of the struggles of a teenager and for an audience composed predominantly of teenagers. His variation thus had more direct impact on male baby-boomer culture than Brando's caveman with the heart of an infant. Dean also brought his own bisexuality to the screen in a way that Brando (although he freely admitted to having had homosexual relationships) hadn't. Although Plato (Sal Mineo's character in *Rebel*, described by him as "the first gay teenager in films"—although still covertly so) seeks a father substitute rather than a lover in Jim Stark (Dean), the physical playfulness and tenderness between them is unlike anything Hollywood had allowed before in a relationship between two men. Screenwriter Stewart Stern wanted that current to be palpable: "One of the things I wanted to show in *Rebel* is that underneath all the bullshit macho defense, there was this pure drive for affection, and it didn't matter who the recipient might be." He was helped by the fact that Dean had affairs with both Sal Mineo and Natalie Wood during the filming of the movie. Unlike Brando, Dean kept his sexual business private; he freely used his own experience, however, to color his screen persona. (That sometimes imaginatively outstripped what one might realistically expect from the characters—as was the case with Jim Stark in *Rebel*, a bit too unthreatened by "the feminine" to ring true as a suburban teenager.)

Brando's Kowalski was also arguably the first screen tough guy to be played with a critical "theory" of the falseness of masculine bravado consciously informing his interpretation. Still other descendants of Brando—Jack Nicholson, Al Pacino, Robert De Niro—picked up on this aspect of Brando's work. Unlike Dean, they didn't bring an androgynous element to their roles, but their interpretation of the aggressively male elements of the characters they played was ironic, slightly detached, bringing out the false bluster and under-the-surface pathos of the characters. Along with directors Mike Nichols, Steven Spielberg, Martin Scorsese, Robert Altman, and Francis Ford Coppola, they made macho and its discontents the subject of critical exposé—not so much in conscious political solidarity with feminism as in cynicism and exhaustion with the cartoonlike, self-destructive social roles that men of their generation were assigned. In their work, variations on Stanley Kowalski in every conceivable ethnic and class incarnation were now appearing on-screen, set in quotation marks, with a question posed at the end of the film. Who is this joker? Is he not a little too dangerous? A bit pathetic?

Kowalski generating anti-Kowalski. Brando's legacy to screen masculinity was rich indeed.

the hidden history of the male sex object

The Legion of Decency knew what it was doing when it chopped up the morning-after scene in which Stella, rumpled and naked in bed, tells Blanche about the lightbulb smashing. "I was sort of thrilled" by it, she admits with a naughty smile to Blanche (who is appalled). It really *is* the most dangerous line, and not only because it is so powerfully suggestive of Stella's pleasure in Stanley's intensity, but because of what it does to the viewer. Evoking the purely exuberant side of Stanley's nature (no nastiness, no violence, just blissful, sexual vitality—Stanley's "animal joy in being," as Williams put it), there's nothing to recoil from and everything to fantasize over. What would it be like to be with a man like *that*? What was one missing by *not* being with a man like that? These are not the sorts of thoughts the Legion of Decency wanted to inspire in female viewers.

It's not only those with a religious agenda, however, who censor, purposefully or not, the sexual subjectivity of women. All feminists know, in theory, that Hollywood rarely plays to the female gaze. But sometimes you only truly realize—viscerally—what's been forbidden when it's finally permitted. Clint Eastwood—who began his career as a "screw you" action hero and wound up directing and starring in one of the best "women's movies" of all time—improved vastly on *The Bridges of Madison County* by telling the story of the affair almost entirely from the heroine's (Meryl Streep) point of view, focusing on her desires, conflicts, and anxieties (rather than the hero's, as was the case with the book). Eastwood's film was the exception that made me aware of what I'd been missing.

From her kitchen window, fanning her face, growing restless and disturbed, Streep watches shirtless Eastwood in the yard (the fact that his chest is frankly aging, sagging, makes the scene all the more erotic; in an era of plastic beauty, Clint was real flesh—and so was Streep, zaftig for the role). Later, in the bathtub, she is transfixed by the drops of water coming from the showerhead that has just recently poured onto his body. That touched you, we imagine her thinking, your body was here, where mine is now. After she's had her bath and dressed—and it's still not clear between them what's going to happen—she places her hand on his shoulder during a phone call. That's all—a hand on his shoulder. But the film has been so finely tuned to her as a sexual subject—except for him telling her she looks "stunning," it never focuses on his reactions to her body, always on her reactions to his—that the erotic charge of the gesture is almost unbearable.

My sister and I, watching the movie, gasped in unison. I felt as though I had put *my* hand on Clint's shoulder. My body was visited by memories of *that* moment in my own life—that moment when a gesture, a look, a meeting of eyes, a brush of skin, makes what's going to happen clear and unalterable. And then I realized what a long time it had been since I'd seen a film which identified with, went deep inside the experience of women's desires. I don't count contemporary television comedies like *Ally McBeal* and *Sex and the City*, whose notions of depicting female desire involve computer-generated serpentlike tongues darting out of their heroine's mouth at the sight of sexy men, or "girls are just as horny as boys" conversations about oral sex and penis size. I mean the world as shown through the medium of our stirring blood, our constricting throats, our quickening pulses.

In 1981, critics squirmed when in George Cukor's *Rich and Famous*, a fresh-faced, very boyish-looking male prostitute (he's either that or simply very generous with his favors) follows a world-weary author (Jacqueline Bisset) through the streets of New York to her hotel room, and proceeds to strip while she watches, sitting on the edge of the bed. As the young man (played by Matt Lattanzi) takes off his shirt, he sucks his stomach in (this was a bit before the washboard era) and preens with pleasure at presenting the sight (and then, the services) of his smooth young body to her. She proceeds to feast at the altar of his beauty and youth. Pauline Kael, in reviewing *Rich and Famous*, snorted that no woman would behave like that, and sniggered that this was a gay movie in disguise. Her homophobic dig at director Cukor (who was gay, as Kael knew very well but couldn't say in her review) also revealed her own discomfort at watching a male depicted so fully and happily as the sex "object" of a woman's desires.

By 1991 and *Thelma & Louise*, male critics were less offended by Brad Pitt as Thelma's (Geena Davis) boy toy than they were by the "male-bashing" stereotypes of husband and truck drivers (or perhaps Pitt's servicing Davis was acceptable because he stole all her money the next morning). But for women, Brad Pitt was as much a break-through element as anything else in that film. It wasn't just how cute he was; it was the fact that his cuteness was *for her*. My female students were jubilant too when that famous Diet Coke commercial aired in 1994. In the commercial, a hunky construction worker (model Lucky Vanos, who became famous for five minutes) takes off his shirt and guzzles a can of Diet Coke, all the while being ogled from their windows by an officeful of women, who had set their watches to the precise time of Vanos's break so that they could watch the show. The women leer while Vanos slowly undresses to strip-show music. The camera lingers on his muscular, nicely hairy chest. Does he know the women are watching him? That's unclear, but in his lavish exhibition of his body there is a slight suggestion that perhaps he does. It was just a dumb commercial, but to many women it felt like the turning of the tide.

It was meant to be seen that way, of course—as a humorous turn-around of the much more familiar scene in which a beautiful woman walks down a street, or enters a room, or is glimpsed at a party, or spied through a window, or is hanging clothes out to dry, or getting change at the supermarket, and rivets the attention of the man (or men) watching her. Michael Corleone in *The Godfather, Part II*, hit by "the thunderbolt" as he glimpses a young Sicilian beauty walking through a field. Burt Lancaster spying on Susan Sarandon putting lemon juice on her arms in *Atlantic City*. Countless male swoons and pratfalls in countless romantic comedies. My father's account of how he met my mother: she came into his family's bakery to buy bread; he took one look at her flaming red hair and lush body and his heart began to race. Did her heart race for him too? That wasn't part of the story. The beautiful female body and the male consciousness that responds to her beauty. It's a motif so familiar that only feminists had noticed how lopsided it was.

Revisiting the films of the fifties, however, gave me some fresh historical perspective on that lopsidedness. Films of the fifties *often* adopted the sexual perspective of a woman looking at the male hero. Blanche nervously eyeing Stanley as he takes off his T-shirt in *Streetcar*. Alexandra De Lago (Geraldine Page) drooling over Chance Wayne (Paul Newman) in *Sweet Bird of Youth*; "I like bodies to be silky smooth, hard, gold," she says as she surveys his (which was indeed, as she notes, "pure, hard gold"). *Everyone* eyeballing Hal Carter's (William Holden) naked chest in *Picnic* (1955). It's fascinating to me that these mid-fifties movies—whose explicit gender politics were hardly the most progressive—so often featured a gorgeous male body as the focus of a female subject's sexual gaze. True, women who looked lasciviously at men were often portrayed in these movies as sex-deprived, voracious, sometimes even villainous (think Barbara Stanwyck). But not always. *Picnic* goes from one female character to another, the young beauties as well as the frustrated spinsters, registering Hal's effect on each. In the process we get to see a lot of Bill Holden shirtless, stripping, or being stripped by others.

The first odd job Hal (a drifter who's just arrived in town via freight train) gets is cleaning kindly Mrs. Potts's yard. Won't he be "awful hot in that jacket"? When Hall tells Potts that his shirt is dirty, she offers to wash it. "Think anyone'll mind?" he asks. Hmmm . . . I wonder. Mrs. Potts (who's the voice of sexual sanity in the movie) says of course not, and Hal does his first strip. Almost immediately, his virile, sweaty body is noticed by frustrated schoolteacher Rosemary (Rosalind Russell), whose eyes register its impact on her, then by the two "heroines" (sisters Madge and Millie, played by Kim Novak and Susan Strasberg) and their straitlaced mother, who live next door. Hal is so embarrassed by the way they're looking at him that he covers his chest like a girl hiding her breasts! It's a ridiculous moment, but establishes that Hal's maleness is going to be a problem for him in this repressed little town.

Indeed it is. The women continue to swoon over his vitality and strength ("He carries that washtub as though it were so much tissue paper!" sighs Mrs. Potts) while Hal's old friend, rich boy Alan Benson (Cliff Robertson), who is engaged to beauty queen Madge, gets jealouser and jealouser. The second time that Hal is obliged to strip occurs when he's borrowing Alan's jacket for the town picnic. It doesn't fit. "I'm beefy through the shoulders," Hal explains. (Thanks, Hal. We hadn't noticed.) So naturally, he has to take it off—and roll his shirt-sleeves up to show off his muscular arms too. None of this is lost on the women, all of whom (except for Mom) have by now developed big crushes on him. ("Absolutely and positively the cutest thing I ever saw," declares one of Madge's girlfriends, as Hal shows off a dive at the swimming pool.)

At the picnic, everyone gets a bit drunk and the bubbling cauldron of sexual feeling overflows. Madge and Hal have a sexy dance together ("Moonglow" sultry in the background), Millie gets sick to her stomach and schoolmarm Rosemary loses her cool entirely, taunting Hal to show his legs and ultimately tearing his shirt. Appalled and humiliated, Hal runs off and Alan sends the cops after him. But Madge catches up with Hal first, and he's obliged to take his shirt off yet again. This time, he rips it off himself in a hokey moment of anger and self-disclosure, as he tells her all about his checkered past (drunkard father, slutty mother, jailed at fourteen for stealing a motorcycle, the usual).

But Madge doesn't care about his past, she loves him because some-how—I'm not sure how—she's gotten the idea that she's more than just a pretty face to him. So they make love and after a few more plot wrinkles, she decides to join him when he's run out of town. (The best scene in the movie is the last one—an aerial shot of her bus traveling across the plains, gradually catching up to the locomotive that is speeding him along.)

Studying the career of the male body in popular culture provides an interesting vantage point from which to reassess the sexual "advances" of the sixties. With a major prohibition on nudity cracked by *The Pawnbroker*, open season was declared on the depiction of women's bodies (as Susan Douglas comments, "the T&A floodgates were opened") and popular culture began to look more and more like a sex-ual cornucopia—for men. At the same time, Hollywood became less and less interested in how women saw and experienced things—as "sub-jects," as it were. At the beginning of the decade, there was *A Summer Place*, *Splendor in the Grass*, *The L-Shaped Room*—all stories of female desire and its (sometimes happy, sometimes sordid) aftermath. But the films with which we most strongly identify the "sexual revolution" of the sixties—*The Graduate*, *Butch Cassidy*, *Blowup*, *8½*, *MASH*—are all told from the perspective of men's crises, desires, development.

It's a paradox, or perhaps just an example of "win some, lose some." The sixties brought women the pill, the Beatles, Helen Gurley Brown, and Gloria Steinem, but drove the male-body-as-sex-object un-derground. William Holden in *Picnic*, Paul Newman in *Sweet Bird of Youth* and *The Long Hot Summer*, Marlon Brando in *A Streetcar Named Desire* were not just the focus of female desire in the *plots* of the films; those plots were exploited to the hilt to allow the camera to lavish erotic attention on the actors' gorgeous faces and bodies. Today, watching those films, I'm struck by how movingly beautiful those men were, what a visual feast it is to look at them. In the sixties and seven-ties, in contrast, as the existential dilemmas of alienated males and the action exploits of male heroes took center stage in the movies, women—now "liberated" from their sexual repression—became the sole focus of the camera's erotic eye, while the "beautiful man" virtu-ally disappeared. I'm not saying, of course, that *Picnic* was a better movie than *MASH*. Or that there weren't still beautiful actors on-screen in the sixties. But far less often were they depicted in a frankly sexualizing way—"as bodies."[*]

the vigorous male animal comes to town

Picnic is an outstanding example of what I'd call the "vigorous male animal comes to town" genre. The heroes in these films—many based on plays by William Inge, William Faulkner, Tennessee Williams—are all strapping, athletic, swaggering, sexually dynamic men who "smell like men" and wear big boots. Often, they have something shady (a false accusation or a real but minor crime) in their pasts. They come to a sleepy midwestern or southern town full of repressed spinster ladies and spirited, yearning-to-be-awakened (but properly brought-up) young heroines, usually with tyrannical fathers or mothers. They wake up the heroines, indeed they do. In fact, they wake everyone up, breathing gusts of fresh, strong male essence wherever they walk.

"*Male.*" As I write these words, I imagine many of my academic colleagues closing this book (or their minds), curling their lips in disgust at my "uninterrogated" use of this term (not to mention my brazen addition of "essence"). But at a certain level, analysis does falter. I look at the publicity stills of Marlon Brando in *Streetcar* and feel myself to be—for whatever complex reasons, biological and/or cultural—among those beings, male and female, who are magnetized and aroused by the maleness of the male body. Call that maleness a myth. Call it a "performance." It's real enough, in our sexual imaginations. Maureen Stapleton, Brando's fellow actor in *Streetcar*, sums it up: "It goes well beyond talent. It's *male.*" The idea is there in *Picnic. Elmer Gantry.*

---

[*]Think of the careers of Warren Beatty and Robert Redford. "I don't mind being an object on the screen, because that's what a role is, something you've created," Redford said in a 1974 interview. "Off screen, I'm not some kind of *thing.* I'm a human being." But when has Redford ever truly been an "object" even on the screen? Even in *The Way We Were*, with Hubbel's WASP good looks a frequent topic of conversation (Katie—Barbra Streisand—calls him her "gorgeous goyishe guy"), the camera lingers far more over Streisand's face than Redford's. In his early films—*The Roman Spring of Mrs. Stone* (1961), *All Fall Down* (1962)—Beatty had a few beefcake shots. But he always seemed ill at ease in them (despite the fact that he looks wonderful). Once he acquired more control of the productions in which he starred, he took his shirt off less and less often. Only in *Shampoo*, where he wears a tight tank top, has Beatty-the-star allowed his body to be on display—and *Shampoo* is a parody, a send-up of narcissism in seventies Beverly Hills.

The Long Hot Summer. Streetcar. Williams may have mourned Stanley's destruction of delicate Blanche, but Stella—who could have gone down the same sad road as Blanche—is saved, her earthiness liberated by his energy. Kowalski, unlike others in the genre I'm describing, is a selfish, wife-beating thug. But he shares with them the essence of a new kind of male protagonist, whose principal role is to release the repressed sexuality of the heroine, by means of his pure, unadulterated maleness.

In the "vigorous male animal comes to town" genre, the hero frequently must first elbow aside another, seemingly more "civilized" (and often richer) man, who doesn't understand the primal needs and desires of the heroine. "He walked in, he clomped through the place like he was still outdoors. There was a man in the place and it seemed good," says Mrs. Potts about Hal Carter in *Picnic*; the implication is that rich boy Alan isn't really a man. In *The Long Hot Summer*, Clara Varner (Joanne Woodward) has been keeping company for many years with a polite member of the landed gentry. When she asks why he's never made a pass, he (also called Alan) tells her that she's "not a girl to be howled at, dragged off into the bushes," but a "nice quiet self-contained girl." Ben Quick (Paul Newman, whose slow walk across the Varners' lawn just about stopped my little adolescent heart) knows better. He knows that "the world belongs to the meat-eaters" (as he tells Clara) and that she is one of them.

"You put those things down, Miss Clara, and I'm gonna kiss you . . . and show you how simple it is." Another line that sizzled its brand on my adolescent consciousness.

The "vigorous male animal comes to town" genre is pure fifties. Since it depends on the existence of a suppressed woman to liberate, it would be an anachronism by the "liberated" sixties. But since it's still the fifties, the sex is not portrayed as entirely "natural" either; there's always that carnal music in the background, for one thing, and other elements signifying sleaze and danger. I was transfixed by *Elmer Gantry* (another "cusp" movie, made in 1960), for its stew of those elements. Gantry (Burt Lancaster) is a vulgar, braggardly, promiscuous salesman, who worms his way into the career, heart, and panties of evangelist Sister Sarah (Jean Simmons). This male animal (unlike Ben Quick, falsely accused of barn burning) really *has* a slimy past. He seduced the preacher's daughter, Lulu, in a pew after the Christmas service, an event described in licentious detail by Lulu (and quoted earlier in this chapter).

Later, Gantry makes good with Sister Sarah on the potency promised by Lulu's description. Gantry has been after Sarah from the start, and while rejecting his overt passes, she's not been entirely discouraging. "Do you think I'm vulgar?" he asks her when they first meet. "Yes," she says, but "you're amusing and you smell like a real man." Uh-oh, churchlady, you're in trouble. Later, he zeroes in for the kill. "I'd like to tear those holy wings off of you and make a real woman out of you," he tells her. "I'd show you what heaven's like . . . ecstasy coming and going." She tries to resist (showing him the tabernacle she's building and insisting this is her only love), but he lures her slowly into the dark of the tent, murmuring soft acknowledgments of everything she's said. "Of course . . . Of course . . ." She follows him into the tent, the music crescendos . . . and the following morning finds her glowing (and humming, of course, which is what we women always do when we are sexually satisfied). At the end, of course, she'll die in a tabernacle fire. But I was only thirteen years old and I was never going to die. Sex was the thing I had to learn about, and clearly I needed a Gantry to teach me.

When I saw these films again recently—a weekend video "chest-fest" that I was, alas, forced to suffer through for this book—the mixed-up feelings that I had about men and men's bodies in those days came flooding back to me. I was convinced that everything thrilling came from them, that they held the deep secret of my sexual being in their power. But the carnal saxophone music was in my brain, and I was frightened too. I rarely knew how to distinguish fear and excitement in myself, or sexual knowingness from adolescent sadism in a boy. In the movies, characters with piercing eyes like Ben Quick's took the upper hand with a woman and told her what she wanted, arrogantly pushing aside her protests, goading her, even taunting and insulting her. I went for boys like that, boys who emanated consciousness of their sexual power—boys who stared, and strutted, and didn't take no for an answer. I thought Ramses-boys (remember Yul in *The Ten Commandments*) had the secret, while "nicer" Moses-boys were clueless. But usually, all Ramses-boys had was the "Ramses"—and sometimes a lot of suppressed hostility for women.

The one I gave myself to did anyway. (He was the one I wrote about earlier, with the "Soft Shoulders" sign.) In high school, he had been the class outlaw and known for his many sexual conquests. He had the stare, the walk, the talk. He also tormented my pet cat and groped my twelve-year-old sister when I wasn't watching. During high school, I lusted madly and ribaldly in my diary ("that body, those hands, that mouth, that smile, those eyes, that walk, that stomach, that prick!" I wrote; I *think* I meant the last as a body part) but only went so far with him. Then I wound up *importing* him to do the job at college—on my eighteenth birthday, in my dorm room. The sex was awful. He hardly touched me, I felt nothing, the condom flew off when it shouldn't have, and I almost got kicked out of the dorm. (We were supposed to keep our doors half open during "parietal hours," as they were called, and I refused to open mine when the monitor knocked.) His treatment of me afterward ran the gamut from cruelty to indifference to occasional invitations to "come up and see him sometime" at his college.

I chose him for the great event over a wonderful young man whom I had begun dating during Christmas break back home. He was the older brother of my best friend and a sophomore at Yale. During the brief time we were involved, he did all kinds of things that charmed and excited me (like stealing a traffic sign for me; in those days, Abbie Hoffman—or some precursor to him—set the standards of romantic etiquette). The problem was, this guy was a Moses, one of the Alans. He had money. He came from a good family. He had lofty ambitions. He was going to be a lawyer. He refused to take advantage of me. (When I got drunk one night and leapt passionately upon him, he very calmly and firmly told me "no," saying he had too much respect for me to do anything while I was in that condition, and that we should wait a little while.) There was nothing of the outlaw about him. His eyes were incapable of piercing anyone's soul. The script for my sexual awakening required someone wild enough to smash lightbulbs; my romantic feeling for him simply withered away.

In real life, however, guys who smashed lightbulbs were, more likely than not, trying to prove their own manhood rather than unleash a girl's womanhood. For as we've seen, the male rebel did not really come out of the jungle to awaken women from their sexual slumber, but to awaken American men to the dangers of becoming tamed, trained, and emasculated by postwar family life. In the movies, these two agendas were presented as blissfully compatible. Ben Quick really *did* know what Clara wanted—in the fiction of the movie—and that turned his harassment (a concept, of course, that we didn't have in those days) into a kind of bold therapy, all aimed at her eventual happiness. Plus, in real life Newman had *married* Joanne Woodward! Joanne Woodward was one actress with whom us less than gorgeous, intelligent types could identify. For many of us, the fact that she had bagged one of the sexiest men who ever lived was definitive reassurance that the happy ending—if not marriage, then at least a long-term relationship—with the dangerous, exciting man was possible.

That never proved true for me. My dangerous first love ultimately joined a Buddhist monastery and spent several years on a mountain somewhere. When he came down, he scolded me for wearing a gauzy peasant skirt through which my underwear could be ever so slightly discerned. Then, after his newfound asceticism had worn off a bit, he slept with my sister. During those days, trying to make sense of the disheveled first stage of my romantic life, I often thought about my friend's brother. Not about how successful he might have become in his career, or how much money he was making—a sixties rebel girl couldn't care less about those things. But I did sometimes wonder what might have happened if I had let a nice boy take his time. I might have actually had some enjoyable sex.

gay men's revenge

the straight man's shadow

As documented in Vito Russo's groundbreaking work, until very recently Hollywood has been a "celluloid closet," in which gay screenwriters and actors have had to keep their own sexuality under wraps and homosexual characters have rarely appeared in films except in coded and disguised form. If you knew your Greek history (or caught the Alan Ladd pinup hanging on the door of his locker) you could "read" the fact that Sal Mineo as Plato in *Rebel without a Cause* is gay. But it was never mentioned, and it was perfectly possible to see him as just a lonely, mixed-up kid.

The specter of effeminacy, however, was lurking behind every corner in the fifties. This was, after all, the decade in which the Marlboro Man was created, after researchers identified Marlboro in 1954 as (horrors) a "feminine high-style cigarette." The reborn campaign, as described by market researcher Pierre Matineau, had a simple formula: "No women in the advertisements . . . All models very virile men . . . Successful, forceful personalities . . . [Above all] reinforce the notion of virility." The original Marlboro man—shown in full glower in this early ad—was bulky ("A lot of man . . . a lot of cigarette") and tattooed, his body a kind of visual masculine hyperbole. To contemporary eyes, he now looks (except for his unsexy bulk) like someone out of a Tom of Finland cartoon, the symbols of masculinity so overdrawn and fetishized as to suggest an ironic, gay sensibility. This itself is the height of irony, since the ad was meant to be determinedly straight—in both senses of the word—in its iconography. The Marlboro Man's excessive virility, in the eyes of the heterosexual admen who created him, was precisely what assured the consumer of his heterosexuality.

Homosexuality as such was not talked about in movies of the fifties. But it was a constant, ghostly presence lurking behind the "problem of masculinity." How much could a guy bend and still be a real man? That issue *was* explicitly being taken up in the movies—in *Rebel*, for example, which resolved it in rather confused fashion. You certainly didn't want to be a feminized househubby, like Jim's dad; the movie was definitive about that. But on the other hand, if you went the Kowalski route, like Buzz—who challenges Jim to a "chicken" race and is unable to get out of his car before it plunges over the cliff—you could end up dead. Judy tells Jim that "a girl wants a man who can be gentle and sweet, and doesn't run away when you need him . . . like

The Marlboro Man

"He gets a lot to like—filter, flavor, flip-top box." The works. A filter that means business. An easy draw that's all flavor. And the flip-top box that ends crushed cigarettes.

being Plato's friend. That's being strong." But would Judy ever have given Jim the time of day if he hadn't proved his courage by taking Buzz's dare? Dependable, brave, sweet, gentle, strong, and independent: maybe James Dean could portray it on-screen, but few real guys (including Dean himself, who met with a fate like Buzz's) could put that package together in life.

Vincente Minnelli's *Tea and Sympathy* (1956, with screenplay by Robert Anderson, who adapted it from his stage play) was another film to pose the question of what makes a man a man, and to challenge some of the old answers. To his father, coach Reynolds, and the other "regular guys" at school, sensitive intellectual Tom Lee (John Kerr) is "different" in all the wrong ways. He knows how to cook and sew. He likes poetry. He doesn't walk right. In one scene, one of the nicer boys, Al, befriends him and tries to teach him how to do it right. He doesn't say exactly what's wrong with the way Tom already walks, but he tries to show Tom—and sort of bounces on his heels across the room. It's not exactly a swish, but it sure isn't the bow-legged, no-nonsense stride that Al demonstrates as the ideal. Tom also spends suspiciously too much time with women (he hadn't gotten Hugh Hefner's separatist message yet, clearly). When he's caught hanging out with a group of faculty wives, sewing a hem, the other guys tease him mercilessly, calling him "sister boy." Other suspect qualities? He has long hair and listens to classical music. And worse yet—as his "regular guy" dad tells the coach, barely able to choke the words out—"he wants to be a . . . folksinger."

Ultimately, Tom achieves manhood and confidence in the warm, sympathetic arms of the coach's wife, Laura (Deborah Kerr), who—like Judy in *Rebel*—argues that "manliness is tenderness, gentleness, consideration," and, when that argument doesn't work, sleeps with him to definitively prove to him that he's a real man. This presumably resolves things for Tom—who at the end of the movie is happily married and a successful author—thus revealing the unstated anxiety that's been lurking behind Tom's insecurities and the other guys' suspicions. (The coach's wife, while confirming Tom's sexual normalcy, got herself into trouble with the Legion of Decency, who insisted that the film end not with her seduction of Tom—as the play had—but with Tom reading a letter from her, ten years later, in which she admits that what she did with him was "wrong" and ruined her husband's life. "There are always consequences," she reminds Tom—a line that could have been written by the Legion itself.)

Within a decade, many of Tom's "problem" traits would no longer be problems, as long hair, listening to classical music, and folksinging became utterly congruent with straight—even aggressively straight—masculinity. By the time of Hal Ashby's *Shampoo* (1975), a promiscuously heterosexual male hero (George, played by Warren Beatty) could even be a hairstylist. *Shampoo* makes fun of those who make the mistake of assuming that George is gay just because he does "heads." Lester (Jack Warden), gangster husband of one of George's rich regulars, Felicia (Lee Grant), makes this assumption and is depicted as out-of-it, a remnant of a less swinging time—and the joke's on him (since George is in fact screwing his wife). But George doesn't just style hair for a living, he cares passionately about it. He sleeps with women, but it's also clear that it's not just about sex. He's much happier and at home in their sensuous, perfumed world than in the male world of bank deals, politics, gray suits.

The film even briefly flirts with the suggestion that George is bisexual. In one scene, Grant's daughter (Carrie Fisher) asks George if he's queer. "Sure," George mumbles. She presses him: "Come on, are you or aren't you?" When he asks why she wants to know and she says it's because she wants to know if he's "making it" with her mother, he asks her what his being homosexual would have to do with that. Neither character ever does answer the other when asked if they've ever had sex with someone of the opposite sex.

But the shadow of Tom Lee was still haunting us, even in *Shampoo*, a sophisticated film consciously aspiring to progressive politics. Just as the possible implications of what George is saying start to register for the viewer, George refers to gays as "faggots"—and it all reconfigures. Beatty (or screenwriter Robert Towne), one now suspects, didn't want George to seem uncool or "insecure about his sexuality" with too swift an assertion of his heterosexuality. But lest we mistake his lack of defensiveness for lack of balls, "faggot" is there to correct things. So, too, George's body helps to keep things straight. Unlike the two men with whom he works in the salon (both of whom are fussy, frivolous types, and dressed in pastels that match the feminine decor of the shop) George appears always in dark clothing, with (nonironic) masculine touches of leather and chaps. In the two scenes in which we see

George having intercourse, he's performing vigorously and unimaginatively, humping away with single-minded dedication in the missionary position. In and out, up and down. Spying on George performing in this way with Jackie (Julie Christie), Lester exclaims admiringly, "Now that's what I call fucking!" Yeah, I guess—if Fred Flintstone is your paradigm. But it wouldn't do to have George be a gentle and sensuous lover; he tends to women's hair by day, he'd better be a no-frills boy by night.

Straight masculinity could only bend so far. In every film in which the hero treads just a little too close to what straight audiences might identify as the gay man's world—*American Gigolo*, for example (1980), in which Richard Gere plays a narcissistic male prostitute—extra insurance is required to make sure that audiences don't get confused. That might mean making the characters ostentatiously heterosexist—in one scene, gigolo Julian Kay (Gere) says he "doesn't do fags." It may also mean trotting out the stereotypes to do some work. So, in another scene, Gere—trying to protect the reputation of his rich "date" when her friend spies them at an antique store—minces, lisps, and swishes, pretending he is her gay male companion. The pretense establishes not only that Julian's "reality" is straight but that his own attitude toward homosexuality is safely condescending. (At the same time, of course, it allows the film to seem hip and bold, letting a straight actor wiggle his butt that way.)

In these films, and many others, the homosexual is invisible yet powerfully *present*—as the shadow of the straight man's sexuality, a constant unseen specter, alluded to through jokes and imitations, the figure against which the heroes must establish their difference. When the homosexual character did appear as a full, flesh-and-blood screen presence, it was as what philosopher Simone de Beauvoir has called "the Other." Unlike straight characters, who get to have exciting adventures in which their sexual orientation is irrelevant, the homosexual character has been continually marked by his or her sexuality. And so too have gay actors—as when Hollywood wonders whether audiences will believe Rupert Everett in straight roles. "Anybody around here ever heard of acting?" columnist Ellen Goodman dryly quipped, commenting on Everett's situation. Heterosexual actors like Robin Williams can play everything from a Russian immigrant to the husband of a drag queen. But gay actors, in the minds of those Hollywood moguls, wear an indelible "GAY" on their foreheads, as do homosexual characters.

Most of the early representations were of slimy underground figures (*Advise & Consent*) and shame-ridden depressives (*Next Stop, Greenwich Village*). Often, the portrayals—even when created by gay screenwriters—traded in stereotypes (as in Mart Crowley's *Boys in the Band*). Later, in the seventies and eighties, the characters became more sympathetic and less caricatured (*Making Love, Torch Song Trilogy*). But with few exceptions, the character's homosexuality and its discontents or complications was a central tenet of the movie. In the nineties, there've been sympathetic dramas about those dying from AIDS. There've been daffy comedies about gays passing for straight. There've been touching coming-out stories. There've been weirdo murderers, the tragically closeted, cynical bitches, flaming sissies, and leather studs. But where were the well-adjusted guys with happy lives and something else to do besides deal with being gay?

"sleek, stylish, and radiant with charisma"

It finally happened in the summer of 1997, and just where you'd least expect it: in a date movie. Rupert Everett as Julianne's—or Jules, as she's usually called (Julia Roberts)—mentor and editor George in *My Best Friend's Wedding* was a first for a major Hollywood production: an openly gay actor playing a gay male who is the moral center of the movie and the best-adjusted, happiest person in it. As urbane as James Bond (but without Bond's misogyny), as homespun-wise as Gramma Walton (but without the sermons), a goofy cutup but also a serious intellectual who spends time at book readings not the gym, George also has all those "caring" virtues usually reserved for female characters. He's tolerant even when Jules interrupts his dinner parties with hysterical phone calls, and drops everything to run to her side when she needs him. But unlike the kindly-gay-male-friend-of-the-heroine as, for example, Harvey Fierstein might play him, it isn't because he hasn't anything better to do. George isn't an ersatz mom or perennially rejected sad sack. God no, this man could get any date he wants, of any sexual orientation.

Even those critics who didn't care for the movie raved about

Everett's performance, describing him as a "charismatic male presence . . . who brazenly swipes every scene that he's in," and predicting
that he would become the "first openly gay sex symbol in Hollywood." Viewers concurred; there's now a thriving "Crazy About Rupert Everett" Internet site, complete with pics, bios, interviews, and
other fan-club paraphernalia. It may be the first openly multisexual fan
club in history, with Everett himself Hollywood's first fully realized
success at what fashion merchandisers refer to as "dual marketing"—
the creation of images which embody elements of strong appeal to different sexual "gazes," gay and straight, male and female. Unlike the
sophisticated fashion world, Hollywood has hit on dual-appeal characters only by chance (or through the covert codings of gay screenwriters and actors), and their appeal has been an underground
phenomenon. Usually, too, it's been gay and lesbian audiences who
have had to eke their pleasures out of slightly "queer" representations
of straight characters: a certain outfit worn by Joan Crawford, a sly
look of Lauren Bacall's, James Dean or Brad Pitt slouching flirtatiously
in T-shirt and jeans.

Everett's George, on the other hand, represented a stolen delight for
straight women. There have, of course, been other gay male characters
in the movies with appeal to women. But George was more than just
appealing; something about him spoke directly—and powerfully—to
women. Unlike the well-hung men in the underwear ads, he doesn't
flaunt his pecs. (He's beautiful to look at, but the closest we get to his
body in the movie is when he wears a sexy black jersey; the rest of the
time he's in well-cut suits.) Yet his sex appeal, as co-star Roberts describes it, was "enormous." "We were dancing in one scene," she recalled in an interview, "and he was being all sexy with me, saying
hilarious things like 'Darling, can't you see it? My last ray of heterosexuality is shining on you.' " Yes, and on the fantasies of every
straight woman who saw the movie.

Those fantasies, moreover, are written right into the movie. At the
very end of the film, when all Jules's schemes to recapture ex-boyfriend
Michael (Dermot Mulroney) have miserably failed and he has married
sugary Kimmy (Cameron Diaz), George makes a surprise appearance
at the wedding reception in an effort to cheer Jules up. He is (as he accurately describes himself) "sleek, stylish, and radiant with charisma."
He rises from his chair, and takes the hand of the beautiful but de-

pressed heroine, pulls her to him, and leads her into a ballroom dance that within minutes has her flushed and bubbling over with delight. He dips and twirls her; they glide across the screen, glamorous and happy, a matched pair of dazzling creatures. Life, as at the end of all comedies, is renewed, refreshed by their being together. For a moment, we might imagine that we are at a revival of screwball comedy, that this is a modern incarnation of Irene Dunne and Cary Grant, a future of (re)wedded bliss ahead of them.

The beautiful heroine doesn't get her man; she gets a dance with her gay male editor. *That* was the happy ending of one of 1997's biggest hits. What did it mean? Reviewers didn't seriously ponder this question, but instead took *My Best Friend's Wedding* at face value, as a somewhat quirky attempt to refresh the "boy meets girl, boy loses girl, boy gets girl" scenario, which departs from the usual formula only in the fact that the boy (in this case a girl) doesn't get her heart's desire in the end. I want to suggest a more culturally informative reading than that. *My Best Friend's Wedding*, as I read it, very deliberately means to "queer" representations of gender and cast doubt on the viability or durability of the fifties version of happily-ever-after. It's not just the character of George that functions this way in the film, but George is essential—not only to challenge stereotypes of homosexuality but also to reveal the inadequacies of *straight* masculinity. This was something critics missed, for all that they raved about Everett, and it was truly something new for Hollywood—a gay male serving up instruction to straight men.

The ideal that Everett/George offered was not an absolutely new one, but a revival of qualities postwar Hollywood in its anxieties about masculinity had buried. Contemporary reviewers have frequently described Everett as a "gay" Cary Grant. None seem to have noticed, as Pauline Kael did in an essay on Grant, that Cary Grant, if only (or perhaps not only) in his screen relations with women, was a bit "gay" himself. Here, Kael contrasts him with Clark Gable:

> With Gable, sex is inevitable: What is there but sex? Basically, he thinks women are good for only one thing. Grant is interested in the qualities of a particular woman—her sappy expression, her non sequiturs, the way her voice bobbles. She isn't going to be pushed to the wall as soon as she's alone with him. With Grant, the social, urban man, there are infinite possibilities for mutual entertainment. They might dance the night away or stroll or go

to a carnival—and nothing sexual would happen unless she wanted it to . . . [I]f the roof leaks, or the car stalls, or you don't know how to get the super to keep his paws off you, you may long for a Clark Gable to take charge, but when you think of going out, Cary Grant is your dream date. . . . How could the heroine ever consider marrying a rich rube from Oklahoma and leaving Cary Grant and the night spots? . . . When [women] look at him, they see a man they could have fun with.

"A man [a woman] could have fun with"—that seems like a pretty superficial appeal. But in the screwball comedies that Grant set his stamp on, having fun is far from a superficial accomplishment. You can't have fun with just anyone, and most men in these films don't really want to; they want to get down to the serious business of finding an "appropriate" mate who will help them to lead respectable, successful lives. As the man a woman could have fun with, Cary Grant is also the man who *wants* to have fun with a woman, who values her "particular" qualities, her independence and vitality, and is uninterested in taming her and turning her into a domestic "helpmate." The rich rube from Oklahoma (a reference to Ralph Bellamy in Leo McCarey's 1937 *The Awful Truth*) is shocked and calls the whole thing off when Lucy Warriner (Irene Dunne) appears to have been carrying on in a less than proper manner. But Jerry Warriner (Gary Grant) smiles admiringly, adoringly at his ex-wife's spirit and nerve, as she mimics the vulgar dance of a singer for the prissy relatives of his new (but not for long) girlfriend.

I didn't realize until I discovered the movies of the thirties and forties that there had been a genre which actually considered playfulness and equality of wit and style between men and women to be a romantic ideal, even an ideal of marriage. That ideal was rooted in class privilege rather than (as in the fifties) gender difference; it didn't matter whether you were a man or a woman in the classic screwball comedy, no one went to work or wore an apron, and everyone had time to fiddle and scheme over their relationships. Cary Grant and Irene Dunne in *The Awful Truth* were a matched set—wisecrack for wisecrack, scheme for scheme, glamour for glamour. They were made for each other not as the positive and negative poles of a magnet, but by virtue of an intimacy and mutual recognition so subtle and implicit that it was slightly incestuous. (Often, the heroes and heroines of screwball comedies have known each other since childhood.) But even such inti-

macy does not come easily and is certainly not guaranteed for the life
of a relationship; it has to be renewed. Dejected by a failure of mutual
understanding, the hero and heroine of the screwball comedy may decide to attempt a life with a more conventional person (e.g., the
"rube"). It can't work, and learning that—learning who one really is
and whom one needs to be with in order to fully realize that—is the
arc of the comedy.

Sex was not the focus of the screwball comedy. Rather, it was the
implicit consequence of intimacy. That the hero and heroine *did* have
sex (never on-screen in those days, but it was often quite clearly alluded to) inspired, in the viewer, the remarkable notion that perhaps
sexuality could be grounded in permission to be fully oneself, with
someone who truly, deeply knows you, and is mad about you all the
same. "I'll be yar, I promise to be yar!" Tracy (Katharine Hepburn)
swears to Dexter (Grant), who's just proposed to marry her again. "Be
whatever you want. You're my redhead," he replies. That one line
(from *The Philadelphia Story*) speaks volumes to the difference between a thirties happily-ever-after and a fifties one, in which everyone,
having ironed out the kinks in their personalities and sowed their wild
oats, settles down and takes their conventional gender places.

The line from *The Philadelphia Story* is one that always brings tears
to my eyes. And, watching the film again recently, I realized that the
moment when Tracy and Dexter recommit was sexually evocative for
me too. Had it always been that way? I doubt it. When I was a
teenager I paid hardly any attention to these movies. By the time I
came to appreciate them, it was the seventies and the cinematic codes
were different. Deep, mutual recognition had dropped out as a prelude
to sex in the movies. And so, in a curious and perverse way (for this
was the era of women's liberation), had a certain ideal of equality as a
precondition of real intimacy. In those days, I expected that love would
result magically from sex, and I certainly didn't require much in the
way of respect or recognition from the one-night stands that were my
laboratories. I suspect my experience was fairly typical.

Rupert Everett's George, like a Cary Grant character, radiates elegance, wit, charm, and playfulness—an importation of the old "class"
values of the screwball comedy into a contemporary Hollywood
movie. And, like Grant in *The Awful Truth* and *The Philadelphia
Story*, his rapport with the movie's heroine is premised on recognition

of who she is and deep affection for her unique qualities—maddening though they may sometimes be. George and Jules understand each other (or, at least, George understands Jules). And—a fact not noted by any reviewers—more touching and nuzzling goes on between George and Jules than between any other characters in the movie, including Michael and Kim. True, some of George's touching of Jules occurs when he is playing the part of her fiancé. But his tenderness for her generates sparks of intimacy nonetheless, not unlike those generated between Cary Grant and Katharine Hepburn at the end of *The Philadelphia Story*.

Audiences believed in these moments of contact, in the feeling behind them. In an earlier version of the movie—a revision of Ronald Bass's original screenplay, demanded by studio bosses who didn't think audiences would accept a gay best friend as compensation for not getting a husband—Jules ends up with a new heterosexual boyfriend. It bombed with preview audiences. The final version deposited the straight boyfriend "on the cutting-room floor" (as Bass has put it). Audiences preferred the specific "chemistry" (or perhaps "alchemy" is a better term) between Roberts and Everett to the conventionally happy ending.

queering the happily-ever-after

Even if we grant that the time was ripe for a reincarnation of Cary Grant and a revival of the screwball comedy, why did Bass and director P. J. Hogan choose to re-create Grant in the persona of a gay man and a friend—not a potential lover—of the heroine? "If this were a vintage screwball comedy [Jules] would have walked off happily ever after on George's exquisitely tailored arm," notes Rick Groen of the *Toronto Globe and Mail*. That she didn't disappointed many reviewers and seemed inexplicable to them. "For some reason," wrote Ruthe Stein of the *San Francisco Chronicle*, "the filmmakers have chosen to make George gay and therefore unavailable. Roberts and Everett are so right for each other," Stein goes on, "it is easy to imagine him as Michael [the longed-for boyfriend]. Now that would have been a movie as irresistible as its star."

Would it? I'm not sure that we *would* believe, in the nineties, in the same gestures as performed by a couple bound for the altar. Our cultural imagination is too cluttered with hundreds of contrived Hollywood happily-ever-afters between Michaels and Kimmys, and marriage as a metaphor for finding one's true self in another just doesn't ring as true as it may have in the thirties. Hogan's first film, *Muriel's Wedding*, certainly took a sardonic view of marriage, and although Hogan makes Michael and Kimmy fairly likable characters in *My Best Friend's Wedding*, he has his tongue firmly in his cheek when it comes to the happily-ever-after part.

Hogan's wry, although not mean-spirited, attitude toward the fifties fantasy is evident from the unmistakably satirical opening credits, featuring a girl group dressed as a bride and her bridesmaids, all sugar-coated and dripping softness, singing "Wishing and Hoping" (the deliberately precious little voice on the soundtrack is actually Ani De-Franco's), to the scene where Michael and Kimmy, newly married, drive off in their car as huge spurting fountains announce their triumph. Viewers may wish and hope that they'll be forever happy, but we know there's trouble down that road.

What reviewers like Ruthe Stein seem to have missed is that George's gayness is an essential ingredient in *My Best Friend's Wedding*'s ironic, deconstructive take on gender. From the opening credits (the lead singer's gestures clearly inspired by Ann-Margret's opening-credit number in *Bye Bye Birdie*), the movie signals its intention to send up the conventional gender roles that are our inheritance from the fifties. George plays an important part in this. Like the parody of the girl group, in his most hilarious scenes he's a fun-house mirror, reflecting in comic form the highly gendered world around him. Asked by Julianne to pretend that he's her ardent, straight fiancé, he grabs at her breast in the cab and shakes male relatives' hands with wrist-breaking gusto ("Just in time for a quick preconjugal visit, if you catch my drift," he winks, man to man). But approached by squealing Kimmy and her mother (who trips daintily down the aisle like a drag queen and who often speaks in the droll bitchy tone of one), he sends *their* "femininity" right back at them: "Love the bags, love the shoes, love everything!" he exclaims.

My Best Friend's Wedding may not want to totally topple the little Barbie/Ken couple from the wedding cake, but it certainly intends that the legacy of the fifties make room (once again) for other kinds of intimacy, other styles of being men and women. In the nineties, that may first require figuring out who one is vis-à-vis a cultural inheritance of gender expectations and sexual roles. Homosexual *or* heterosexual, some people are "queer"—they don't fit easily into those pigeonholes. Jules, although played by a gorgeous Julia Roberts, is much less a Pretty Woman and more like a typical male from the self-help guides written for frustrated wives and girlfriends. Jules can't commit. A workaholic who doesn't like to smooch in public (when people embrace her, she cringes a bit) and who has cried only three times in her life, she was too self-contained for Michael, who wants someone more cuddly—which Kimmy, always on the verge of tears, giggles, or a hug, certainly is. Until she is provoked by the prospect of losing Michael to Kimmy, Jules hasn't been interested in "female priorities" like love and marriage; she can't stand "yucky love stuff."

Kimmy, on the other hand, is so stereotypically feminine that she's virtually a whole other gender. (Except for her frail blondness, Kimmy could be Elizabeth Taylor in *Father of the Bride*.) When Jules likens Kimmy to crème brûlée, suggesting that Michael might be more of a Jell-O man, Kimmy replies tearfully, "*I* can be Jell-O! I *have* to be Jell-O!" This is another moment when it's hard to believe that Bass and Hogan want us to admire Kimmy as wholeheartedly as reviewers seem to have thought. (Reviewers described her as "a spunky, ingenious, kindly spirit," "dainty, friendly," "goodness and warmth," "delightfully daffy," and so on.) But more important than whether Kimmy is "smart and spunky" or a sap is the fact that whatever her virtues, they are utterly different from Julianne's. When Kim wins over a karaoke bar with her off-key, tremulous rendition of (appropriately enough), "I just don't know what to do with myself," it's clear from Jules's astonished face, watching Kimmy perform, that she realizes she is in the presence of something utterly and bafflingly "other" to herself.

Although she sings that she doesn't, in actual fact Kimmy *does* know what to do with herself; it's Jules who doesn't. Jules is determined to become the girl on top of the wedding cake, even if it means resisting the truth of her own personality. The movie, of course, won't let her, and George is integral to that resolution, not just because he is the one who gets Jules to "do the right thing" and give up Michael, but because he shows us—by being "Cary Grant," the man a girl would *re-*

ally like to play with—just how boring a fifties-style happily-ever-after would be for high-spirited, independent Julianne. "It would never have worked out" with Michael, George tells Jules. "Different tempera-ments." But it's really George's "radiant charisma" and not his dis-pensed wisdom that disabuses *us*—the viewers—of any misguided notion we may have that Jules would be happy in a conventional mat-ing with a conventional man. Kim and Michael—who even look like a little toy couple when they're side by side—do indeed belong together on top of the wedding cake. But not everyone is as comfortably fitted in his or her gender skin as Michael and Kimmy. In a world that's or-ganized for those who are, it's a painful recognition to realize that one isn't—and Jules struggles against it.

George is proof, however, that letting go of gender rules has its com-pensations. Like Jules in the karaoke bar, conventional Michael has a moment of self-recognition when George's sultry, comic rendition of "Say a Little Prayer for You" transforms an entire restaurant into the Supremes. At that moment Michael begins to suspect that he, Michael, is the rube from Oklahoma (so to speak)—a nice, square dude but ut-terly without George's exuberant charm. Hollywood's obsessive exor-cism of the shadow of homosexuality has finally caught up with the straight male hero, revealing to him the boring bland character that he has become. Other scenes too show that being queer can be fun. Be-fore the wedding, Michael's little brother and a few other young guys (one with an earring in his ear), blowing up helium balloons, perform a chipmunk version of "You Fill Up My Senses." They don't crack up, as we might expect from a bunch of young boys, but close their eyes dreamily, sweetly crooning together, enjoying the chance to play at be-ing something other than "guys." Their rendition is touching, an ab-surd moment turned luminous for the viewer—as the restaurant scene is too—by the uninhibited willingness of the characters to step outside the norms of behavior, to expand rather than to contract, to go for eros, even in unconventional forms.

And then there's the dance at the end of the movie. Early in *My Best Friend's Wedding*, we're told that Jules can't dance. It is Michael who tells us. "You can't dance! When did you learn how to dance?" he bursts out when Jules mentions that the best man (Michael's little brother) must dance with the maid of honor. "I've got moves you haven't seen . . ." Jules replies, and wiggles playfully. The moment strongly recalls the moment in *The Awful Truth* in which Dan, the rube, tells Jerry that Lucy "doesn't care much about dancing." But Jerry knows better, and watches smiling as Dan drags Lucy onto the dance floor and forces her into an unromantic, inelegant jitterbug. From Jerry's self-satisfied face as he watches them dance, we know he knows that Dan will never last. For Lucy *can* dance ("We used to call her twinkle-toes," says Jerry); she just needs the right man to dance with.

So, it appears, does Jules. But in 1998, perhaps, a girl can't have everything.* As George takes her hand and pulls her to him, in a dance that is no less romantic for the fact that they twirl each other by turns, he admits that there are limits to their relationship. "Maybe there won't be marriage, maybe there won't be sex," he admits, exaggerating a grimace, "but by God, there will be dancing!" Thousands of straight women watching him found this a pretty good bargain—a just and fitting revenge for the gay male's Hollywood career as the shadow of straight male sexuality.

---

*Since *My Best Friend's Wedding*, the gay male/straight female relationship has become something of a cultural paradigm, as other movies—*Object of My Affection*—and even prime-time television—*Will and Grace*—play with its possibilities for new models of male-female intimacy.

beauty (re)discovers the male body

men on display

Putting classical art to the side for the moment, the naked and near-naked female body became an object of mainstream consumption first in *Playboy* and its imitators, then in movies, and only then in fashion photographs. With the male body, the trajectory has been different. Fashion has taken the lead, the movies have followed. Hollywood may have been a chest-fest in the fifties, but it was male clothing designers who went south and violated the really powerful taboos—not just against the explicit depiction of penises and male bottoms but against the admission of all sorts of forbidden "feminine" qualities into mainstream conceptions of manliness.

It was the spring of 1995, and I was sipping my first cup of morning coffee, not yet fully awake, flipping through *The New York Times Magazine*, when I had my first real taste of what it's like to inhabit this visual culture as a man. It was both thrilling and disconcerting. It was the first time in my experience that I had encountered a commercial representation of a male body that seemed to deliberately invite me to linger over it. Let me make that stronger—that seemed to reach out to me, interrupting my mundane but peaceful Sunday morning, and provoke me into erotic consciousness, whether or not I wanted it. Women—both straight and gay—have always gazed covertly, of course, squeezing our illicit little titillations out of representations designed for—or pretending to—other purposes than to turn us on. *This*

Calvin Klein

ad made no such pretense. It caused me to knock over my coffee cup, ruining the more cerebral pleasures of the *Book Review*. Later, when I had regained my equilibrium, I made a screen-saver out of him, so I could gaze at my leisure.

I'm sure that many gay men were as taken as I was, and perhaps some gay women too. The erotic charge of various sexual styles is not neatly mapped onto sexual orientation (let alone biological sex). Brad Pitt's baby-butch looks are a turn-on to many lesbians, while I—regarded by most of my gay friends as a pretty hard-core heterosexual—have always found Anne Heche irresistible (even before Ellen did). A lesbian friend of mine, reading a draft of my section on biblical S&M, said the same movies influenced her later attraction to butch *women*. Despite such complications, until recently only heterosexual men have continually been inundated by popular cultural images *designed* with their sexual responses (or, at least, what those sexual responses are imagined to be) in mind. It's not entirely a gift. On the minus side is having one's composure continually challenged by what Timothy Beneke has aptly described as a culture of "intrusive images," eliciting fantasies, emotions, and erections at times and in places where they might not be appropriate. On the plus side is the cultural permission to be a voyeur.

Some psychologists say that the circuit from eyes to brain to genitals is a quicker trip for men than for women. "There's some strong evidence," popular science writer Deborah Blum reports, citing studies of men's responses to pictures of naked women, "that testosterone is wired for visual response." Maybe. But who is the electrician here? God? Mother Nature? Or Hugh Hefner? Practice makes perfect. And women have had little practice. The Calvin Klein ad made me feel like an adolescent again, brought me back to that day when I saw Barry Resnick on the basketball court of Weequahic High and realized that men's legs could make me weak in the knees. Men's legs? I knew that *women's* legs were supposed to be sexy. I had learned that from all those hose-straightening scenes in the movies. But men's legs? Who had ever seen a woman gaga over some guy's legs in the movies? Or even read about it in a book? Yet the muscular grace of Barry's legs took my breath away. Maybe something was wrong with me. Maybe my sex drive was too strong, too much like a man's. By the time I came across that Calvin Klein ad, several decades of feminism and life experience had left me a little less worried about my sex drive. Still, the sight of that model's body made me feel that my sexual education was still far from complete.

I brought the ad to classes and lectures, asking women what they thought of him. Most began to sweat the moment I unfolded the picture, then got their bearings and tried to explore the bewitching stew of sexual elements the picture has to offer. The model—a young Jackson Browne look-alike—stands there in his form-fitting and rip-speckled Calvin Klein briefs, head lowered, dark hair loosely falling over his eyes. His body projects strength, solidity; he's no male waif. But his finely muscled chest is not so overdeveloped as to suggest a sexuality immobilized by the thick matter of the body. Gay theorist Ron Long, describing contemporary gay sexual aesthetics—lean, taut, sinuous muscles rather than Schwarzenegger bulk—points to a "dynamic tension" that the incredible hulks lack. Stiff, engorged Schwarzenegger bodies, he says, seem to *be* surrogate penises—with nowhere to go and nothing to do but stand there looking massive—whereas muscles like this young man's seem designed for movement, for sex. His body isn't a stand-in phallus; rather, he *has* a penis—the real thing, not a symbol, and a fairly breathtaking one, clearly outlined through the soft jersey fabric of the briefs. It seems slightly erect, or perhaps that's his non-erect size; either way, there's a substantial presence there that's palpable (it looks so touchable, you want to cup your hand over it) and very, very male.

At the same time, however, my gaze is invited by something "feminine" about the young man. His underwear may be ripped, but ever so slightly, subtly; unlike the original ripped-underwear poster boy Kowalski, he's hardly a thug. He doesn't stare at the viewer challengingly, belligerently, as do so many models in other ads for male underwear, facing off like a street tough passing a member of the rival gang on the street. ("Yeah, this is an underwear ad and I'm half naked. But I'm still the one in charge here. Who's gonna look away first?") No, this model's languid body posture, his averted look are classic signals, both in the "natural" and the "cultural" world, of willing subordination. He offers himself nonaggressively to the gaze of another. Hip cocked in the snaky S-curve usually reserved for depictions of women's bodies, eyes downcast but not closed, he gives off a sultry, moody, subtle but undeniably seductive consciousness of his erotic allure. Feast on me, I'm here to be looked at, my body is for your eyes. Oh my.

Such an attitude of male sexual supplication, although it has (as we'll see) classical antecedents, is very new to contemporary mainstream representations. Homophobia is at work in this taboo, but so are attitudes about gender that cut across sexual orientation. For many men, both gay and straight, to be so passively dependent on the gaze of another person for one's sense of self-worth is incompatible with being a real man. As we'll see, such notions about manliness are embedded in Greek culture, in contemporary visual representation, and even (in disguised form) in existentialist philosophy. "For the woman," as philosopher Simone de Beauvoir writes, ". . . the absence of her lover is always torture; he is an eye, a judge . . . away from him, she is dispossessed, at once of herself and of the world." For Beauvoir's sometime lover and lifelong soul mate Jean-Paul Sartre, on the other hand, the gaze (or the Look, as he called it) of another person—including the gaze of one's lover—is the "hell" that other people represent. If we were alone in the world, he argues, we would be utterly free—within physical constraints—to be whomever we wanted to be, to be the creatures of our own self-fantasies, to define our behavior however we like. Other people intrude on this solipsism, and have the audacity to see us from their own perspective rather than ours. The result is what Sartre calls primordial Shame under the eyes of the Other, and a fierce desire to reassert one's freedom. The other person has stolen "the secret" of who I am. I must fight back, resist their attempts to define me.

I understand, of course, what Sartre is talking about here. We've all, male and female alike, felt the shame that another pair of eyes can bring. Sartre's own classic example is of being caught peeking through a keyhole by another person. It isn't until those other eyes are upon you that you truly feel not just the "wrongness" of what you are doing, but—Sartre would argue—the very fact that you are doing it. Until the eyes of another are upon us, "catching us" in the act, we can deceive ourselves, pretend. Getting caught in moments of fantasy or vanity may be especially shameful. When I was an adolescent, I loved to pretend I was a radio personality, and talking into an empty coffee can created just the right sound. One day, my mother caught me speaking in the smooth and slightly sultry tones that radio personalities had even in those days. The way I felt is what Sartre means when he describes the Look of another person as the fulcrum of shame-making. My face got hot, and suddenly I saw how ridiculous I must have seemed, my head in the Chock Full O' Nuts, my narcissistic fantasies on full display. I was caught, I wanted to run.

The disjunction between self-conception and external judgment can be especially harsh when the external definitions carry racial and gender stereotypes with them. Sartre doesn't present such examples—he's interested in capturing the contours of an existential situation shared by all rather than in analyzing the cultural differences that affect that situation—but they are surely relevant to understanding the meaning of the Look of the Other. A black man jogs down the street in sweat clothes, thinking of the class he is going to teach later that day; a white woman passes him, clutches her handbag more tightly, quickens her step; in her eyes, the teacher is a potentially dangerous animal. A Latin American student arrives early the first day of college; an administrator, seeing him in the still-deserted hall, asks him if he is the new janitor. The aspiring student has had his emerging identity erased, a stereotype put in its place by another pair of eyes. When women are transformed from professionals to "pussies" by the comments of men on the street, it's humiliating, not so much because we're puritans as because we sense the hostility in the hoots, the desire to bring an uppity woman down to size by reminding her that she's just "the sex" (as Beauvoir put it).

We may all have felt shame, but—as the different attitudes of Beauvoir and Sartre suggest—men and women are socially sanctioned to deal with the gaze of the Other in different ways. Women learn to anticipate, even play to the sexualizing gaze, trying to become what will please, captivate, turn shame into pride. (In the process, we also learn how sexy being gazed at can feel—perhaps precisely because it walks the fine edge of shame.) Many of us, truth be told, get somewhat addicted to the experience. I'm renting a video, feeling a bit low, a bit tired. The young man at the counter, unsolicited, tells me I'm "looking good." It alters everything, I feel fine, alive; it seems to go right down to my cells. I leave the store feeling younger, stronger, more awake. When women sense that they are not being assessed sexually—for example, as we age, or if we are disabled—it may feel like we no longer exist.

Women may dread being surveyed harshly—being seen as too old, too fat, too flat-chested—but men are not supposed to enjoy being surveyed *period*. It's feminine to be on display. Men are thus taught—as my uncle Leon used to say—to be a moving target. Get out of range of those eyes, don't let them catch you—even as the object of their fantasies (or, as Sartre would put it, don't let them "possess," "steal" your freedom)." This phobia has even distorted scientific research, as mentioned earlier. Evolutionary theorists have long acknowledged display as an important feature of courting behavior among primates—except when it comes to *our* closest ancestors. With descriptions of hominid behavior, male display behavior "suddenly drops out of the primate evolutionary picture" (Sheets-Johnstone) and is replaced by the concept of year-round female sexual receptivity. It seems that it has been intolerable, unthinkable for male evolutionary theorists to imagine the bodies of their male ancestors being on display, sized up, dependent on selection (or rejection) by female hominids.

Scientists and "ordinary guys" are totally in synch here, as is humorously illustrated in Peter Cattaneo's popular 1997 British film *The Full Monty*. In the film, a group of unemployed metalworkers in Sheffield, England, watch a Chippendale's show and hatch the money-making scheme of presenting their own male strip show in which they will go right down to the "full Monty." At the start of the film, the heroes are hardly pillars of successful manliness (Gaz, their leader, refers to them as "scrap"). Yet even they have been sheltered by their guy-hood, as they learn while putting the show together. One gets a penis pump. Another borrows his wife's face cream. They run, they wrap their bellies in plastic, they do jumping jacks, they get artificial tans. The most overweight one among them (temporarily) pulls out of the show. Before, these guys hadn't lived their lives under physical scrutiny, but in male action mode, in which men are judged by their accomplishments. Now, anticipating being on display to a roomful of spectators, they suddenly realize how it feels to be judged as women routinely are, sized up by another pair of eyes. "I pray that they'll be a bit more understanding about us" than they've been with women, David (the fat one) murmurs.

They get past their discomfort, in the end, and their show is greeted with wild enthusiasm by the audience. The movie leaves us with this feel-good ending, not raising the question obvious to every woman watching the film: would a troupe of out-of-shape women be received as warmly, as affectionately? The climactic moment when the men throw off their little pouches is demurely shot from the rear, moreover, so we—the audience—don't get "the full Monty." Nonetheless, the film gently and humorously makes an important point: for a hetero-sexual man to offer himself up to a sexually evaluating gaze is for him to make a large, scary leap—and not just because of the anxieties about size discussed earlier in this book (the guy who drops out of the show, remember, is embarrassed by his fat, not his penis). The "full Monty"—the naked penis—is not merely a body part in the movie (hence it doesn't really matter that the film doesn't show it). It's a symbol for male exposure, vulnerability to an evaluation and judgment that women—clothed or naked—experience all the time.

I had to laugh out loud at a 1997 *New York Times Magazine* "Style" column, entitled "Overexposure," which complained of the "contagion" of nudity spreading through celebrity culture. "Stars no longer have private parts," the author observed, and fretted that civilians would soon also be measured by the beauty of their buns. I share this author's concern about our body-obsessed culture. But, pardon me, he's just noticing this now??? Actresses have been baring their breasts, their butts, even their bushes, for some time, and ordinary women have been tromping off to the gym in pursuit of comparably perfect bodies. What's got the author suddenly crying "overkill," it turns out, is Sly Stallone's "surreally fat-free" appearance on the cover of *Vanity Fair*, and Rupert Everett's "dimpled behind" in a Karl Lagerfeld fashion spread. Now that *men* are taking off their clothes, the culture is suddenly going too far. Could it be that the author doesn't even "read" all those naked female bodies as "overexposed"? Does he protest a bit too much when he declares in the first sentence of the piece that he found it "a yawn" when Dirk Diggler unsheathed his "prosthetic shillelagh" ("penis" is still a word to be avoided whenever possible) at the end of *Boogie Nights*? A yawn? My friend's palms were sweating profusely, and I was not about to drop off to sleep either.

As for dimpled behinds, my second choice for male pinup of the decade is the Gucci series of two ads in which a beautiful young man, shot from the rear, puts on a pair of briefs. In the first ad, he's holding them in his hands, contemplating them. Is he checking out the correct washing-machine temp? It's odd, surely, to stand there looking at your underwear, but never mind. The point is: his underwear is in his hands, not on his butt. *It*—his bottom, that is—is gorgeously, completely naked—a motif so new to mainstream advertising (but since then catching on rapidly) that several of my friends, knowing I was writing

about the male body, E-mailed me immediately when they saw the ad. In the second ad, he's put the underwear on, and is adjusting it to fit. Luckily for us, he hasn't succeeded yet, so his buns are peeking out the bottom of the underwear, looking biteable. For the *Times* writer, those buns may be an indecent exposure of parts that should be kept private (or they're a boring yawn, I'm afraid he can't have it both ways), but for me—and for thousands of gay men across the country—this was a moment of political magnitude, and a delicious one. The body parts that *we* love to squeeze (those plastic breasts, they're the real yawn for me) had come out of the closet and into mainstream culture, where *we* can enjoy them without a trip to a specialty store.

But all this is very new. Women aren't used to seeing naked men frankly portrayed as "objects" of a sexual gaze (and neither are het-erosexual men, as that *Times* writer makes clear). So pardon me if I'm skeptical when I read arguments about men's greater "biological" re-sponsiveness to visual stimuli. These "findings," besides being ethno-centric (no one thinks to poll Trobriand Islanders), display little awareness of the impact of changes in cultural representations on our capacities for sexual response. Popular science writer Deborah Blum, for example, cites a study from the Kinsey Institute which showed a group of men and women a series of photos and drawings of nudes, both male and female:

> Fifty-four percent of the men were erotically aroused versus 12 percent of the women—in other words, more than four times as many men. The same gap exists, on a much larger scale, in the business of pornography, a $500-million-plus industry in the U.S. which caters almost exclusively to men. In the first flush of 1970s feminism, two magazines—*Playgirl* and *Viva*—began publishing male centerfolds. *Viva* dropped the nude photos after surveys showed their readers didn't care for them; the editor herself admitted to find-ing them slightly disgusting.

Blum presents these findings as suggestive of a hard-wired difference between men and women. I'd be cautious about accepting that conclu-sion. First of all, there's the question of which physiological responses count as "erotic arousal" and whether they couldn't be evidence of other states. Clearly, too, we can *learn* to have certain physiological re-sponses—and to suppress them—so nothing biologically definitive is proved by the presence or absence of physical arousal.

Studies that rely on viewers' *own* reports need to be carefully interpreted too. I know, from talking to women students, that they sometimes aren't all that clear about *what* they feel in the presence of erotic stimuli, and even when they are, they may not be all that comfortable admitting what they feel. Hell, not just my students! Once, a lover asked me, as we were about to part for the evening, if there was anything that we hadn't done that I'd really like to do. I knew immediately what that was: I wanted him to undress, very slowly, while I sat on the floor and just watched. But I couldn't tell him. I was too embarrassed. Later, alone in my compartment on the train, I sorely regretted my cowardice. The fact is that I love to watch a man getting undressed, and I especially like it if he is conscious of being looked at. But there is a long legacy of shame to be overcome here, for both sexes, and the cultural models are only now just emerging which might help us move beyond it.

Perhaps, then, we should wait a bit longer, do a few more studies, before we come to any biological conclusions about women's failure to get aroused by naked pictures. A newer (1994) University of Chicago study found that 30 percent of women ages eighteen to forty-four and 19 percent of women ages forty-five to fifty-nine said they found "watching a partner undress" to be "very appealing." ("Not a bad percentage," Nancy Friday comments, "given that Nice Girls didn't look.") There's still a gender gap—the respective figures for men of the same age groups were 50 percent and 40 percent. We're just learning, after all, to be voyeuses. Perhaps, too, heterosexual men could learn to be less uncomfortable offering themselves as "sexual objects" if they realized the pleasure women get from it. Getting what you have been most deprived of is the best gift, the most healing gift, the most potentially transforming gift—because it has the capacity to make one more whole. Women have been deprived not so much of the *sight* of beautiful male bodies as the experience of having the male body *offered* to us, handed to us on a silver platter, the way female bodies—in the ads, in the movies—are handed to men. Getting this from her partner is the erotic equivalent of a woman's coming home from work to find a meal prepared and ready for her. Delicious—even if it's just franks and beans.

thanks, calvin!

Despite their bisexual appeal, the cultural genealogy of the ads I've been discussing and others like them is to be traced largely through gay male aesthetics, rather than a sudden blossoming of appreciation for the fact that women might enjoy looking at sexy, well-hung young men who don't appear to be about to rape them. Feminists might like to imagine that Madison Avenue heard our pleas for sexual equality and finally gave us "men as sex objects." But what's really happened is that women have been the beneficiaries of what might be described as a triumph of pure consumerism—and with it, a burgeoning male fitness and beauty culture—over homophobia and the taboos against male vanity, male "femininity," and erotic display of the male body that have gone along with it.

Throughout this century, gay photographers have created a rich, sensuous, and dramatic tradition which is unabashed in eroticizing the male body, male sensuousness, and male potency, including penises. But until recently, such representations have been kept largely in the closet. Mainstream responses to several important exhibits which opened in the seventies—featuring the groundbreaking early work of Wilhelm von Gloeden, George Dureau, and George Platt Lynes as well as then-contemporary artists such as Robert Mapplethorpe, Peter Hujar, and Arthur Tress—would today probably embarrass the critics who wrote about them when they opened. John Ashbery, in *New York* magazine, dismissed the entire genre of male nude photography with the same sexist tautology that covertly underlies that *Times* piece on cultural "overexposure": "Nude women seem to be in their natural state; men, for some reason, merely look undressed . . . When is a nude not a nude? When it is male." (Substitute "blacks" and "whites" for "women" and "men" and you'll see how offensive the statement is.)

For other reviewers, the naked male, far from seeming "merely undressed," was unnervingly sexual. *New York Times* critic Gene Thompson wrote that "there is something disconcerting about the sight of a man's naked body being presented as a sexual object"; he went on to describe the world of homoerotic photography as one "closed to most of us, fortunately." Vicki Goldberg, writing for the *Saturday Review*, was more appreciative of the "beauty and dignity" of the nude male body, but concluded that so long as its depiction was erotic in emphasis, it will "remain half-private, slightly awkward, an art form cast from its traditions and in search of some niche to call its home."

Goldberg needed a course in art history. It's true that in classical art, the naked human body was often presented as a messenger of spiritual themes, and received as such. But the male bodies sculpted by the Greeks and Michelangelo were not exactly nonerotic. It might be more accurate to say that in modernity, with the spiritual interpretation of the nude body no longer a convention, the contemporary homophobic psyche is not screened from the sexual charge of the nude male body. Goldberg was dead wrong about something else too. Whatever its historical lineage, the frankly sexual representation of the male body was to find, in the next twenty years, a far from private "niche to call its home": consumer culture discovered its commercial potency.

Calvin Klein had his epiphany, according to one biography, one night in 1974 in New York's gay Flamingo bar:

> As Calvin wandered through the crowd at the Flamingo, the body heat rushed through him like a revelation; this was the cutting edge. . . . [The] men! The men at the Flamingo had less to do about sex for him than the notion of portraying men as gods. He realized that what he was watching was the freedom of a new generation, unashamed, in-the-flesh embodiments of Calvin's ideals: straight-looking, masculine men, with chiseled bodies, young Greek gods come to life. The vision of shirtless young men with hardened torsos, all in blue jeans, top button opened, a whisper of hair from the belly button disappearing into the denim pants, would inspire and inform the next ten years of Calvin Klein's print and television advertisements.

Klein's genius was that of a cultural Geiger counter; his own bisexuality enabled him to see that the phallic body, as much as any female figure, is an enduring sex object within Western culture. In America in 1974, however, that ideal was still largely closeted. Only gay culture unashamedly sexualized the lean, fit body that virtually everyone, gay and straight, now aspires to. Sex, as Calvin Klein knew, sells. He also knew that gay sex wouldn't sell to straight men. But the rock-hard, athletic gay male bodies that Klein admired at the Flamingo did not advertise their sexual preference through the feminine codes—limp wrists, raised pinkie finger, swishy walk—which the straight world then identified with homosexuality. Rather, they embodied a highly masculine aesthetic that—although definitely exciting for gay men— would scream "heterosexual" to (clueless) straights. Klein knew just the kind of clothing to show that body off in too. As Steven Gaines and Sharon Churcher tell it:

> He had watched enough attractive young people with good bodies in tight jeans dancing at the Flamingo and Studio 54 to know that the "basket" and the behind was what gave jeans sex appeal. Calvin sent his assistants out for several pairs of jeans, including the classic five-button Levi's, and cut them apart to see how they were made. Then he cut the "rise," or area from the waistband to under the groin, much shorter to accentuate the crotch and pull the seam up between the buttocks, giving the behind more shape and prominence. The result was instant sex appeal—and a look that somehow Calvin just *knew* was going to sell.

So we come to the mainstream commercialization of the aesthetic legacy of Stanley Kowalski and those inspired innovations of Brando's costumer in *A Streetcar Named Desire*. When I was growing up, jeans were "dungarees"—suitable for little kids, hayseeds, and juvenile delinquents, but not for anyone to wear on a date. Klein transformed jeans from utilitarian garments to erotic second skins. Next, Klein went for underwear. He wasn't the first, but he was the most daring. In 1981, Jockey International had broken ground by photographing Baltimore Oriole pitcher Jim Palmer in a pair of briefs (airbrushed) in one of its ads—selling $100 million worth of underwear by year's end. Inspired by Jockey's success, in 1983 Calvin Klein put a forty-by-fifty-foot Bruce Weber photograph of Olympic pole vaulter Tom Hintinauss in Times Square, Hintinauss's large penis clearly discernible through his briefs. The Hintinauss ad, unlike the Palmer ad, did not employ any of the usual fictional rationales for a man's being in his underwear—for example, the pretense that the man is in the process of getting dressed—but blatantly put Hintinauss's body on display, sunbathing on a rooftop, his skin glistening. The line of shorts "flew off the shelves" at Bloomingdale's and when Klein papered bus shelters in Manhattan with poster versions of the ad they were all stolen overnight.

Images of masculinity that will do double (or triple or quadruple) duty with a variety of consumers, straight and gay, male and female, are not difficult to create in a culture like ours, in which the muscular

Bronzed and beautiful Tom Hintinauss: a breakthrough ad for Calvin Klein—and the beginning
of a new era for the unabashed erotic display of the male body

male body has a long and glorious aesthetic history. That's precisely
what Calvin Klein was the first to recognize and exploit—the possibility and profitability of what is known in the trade as a "dual marketing" approach. Since then, many advertisers have taken advantage of

Klein's insight. A recent Abercrombie & Fitch ad, for example, depicts a locker room full of young, half-clothed football players getting a postmortem from their coach after a game. Beautiful, undressed male bodies doing what real men are "supposed to do." Dirty uniforms and smudged faces, wounded players, helmets. What could be more straight? But as iconography depicting a culture of exclusively male bodies, young, gorgeous, and well-hung, what could be more "gay"?

It required a Calvin Klein to give the new vision cultural form. But the fact is that if we've entered a brave, new world of male bodies it is largely because of a more "material" kind of epiphany—a dawning recognition among advertisers of the buying power of gay men. For a long time prejudice had triumphed over the profit motive, blinding marketers to just how sizable—and well-heeled—a consumer group gay men represent. (This has been the case with other "minorities" too. Hollywood producers, never bothering to do any demographics on middle-class and professional African-American women—or the issues that they share with women of other races and classes in this culture—were shocked at the tremendous box office success of *Waiting to Exhale*. They won't make that particular mistake again.) It took a survey conducted by *The Advocate* to jolt corporate America awake about gay consumers. The survey, done between 1977 and 1980, showed that 70 percent of its readers aged twenty to forty earned incomes well above the national median. Soon, articles were appearing on the business pages of newspapers, like one in 1982 in *The New York Times Magazine*, which described advertisers as newly interested in "wooing . . . the white, single, well-educated, well-paid man who happens to be homosexual."

"Happens to be homosexual": the phrasing—suggesting that sexual identity is peripheral, even accidental—is telling. Because of homophobia, dual marketing used to require a delicate balancing act, as advertisers tried to speak to gays "in a way that the straight consumer will not notice." Often, that's been accomplished through the use of play and parody, as in Versace's droll portraits of men being groomed and tended by male servants, and Diesel's overtly narcissistic gay posers. "Thanks, Diesel, for making us so very beautiful," they gush. Or take the ad on the following page, with its gorgeous, mechanically inept model admitting that he's "known more for my superb bone construction and soft, supple hair than my keen intellect." The playful tone

reassures heterosexual consumers that the vanity (and mechanical incompetence) of the man selling the product is "just a joke." For gay
consumers, on the other hand, this reassurance is *itself* the "joke";
they read the humor in the ad as an insider wink, which says, "This is
for *you*, guys." The joke is further layered by the fact that they know
the model in the ad is very likely to be gay.

Contrast this ad to the ostentatious heterosexual protest of a Perry
Ellis ad which appeared in the early 1990s (and no, it's not a parody):

> I hate this job. I'm not just an empty suit who stands in front of a camera,
> collects the money and flies off to St. Maarten for the weekend.
>
> I may model for a living, but I hate being treated like a piece of meat. I
> once had a loud-mouthed art director say "Stand there and pretend you're a
> human." I wanted to punch him, but I needed the job.
>
> What am I all about? Well, I know I'm very good-looking, and there are
> days when that is enough. Some nights, when I'm alone, it's not.

I like women—all kinds.

I like music—all kinds.

I like myself so I don't do drugs.

Oh yeah, about this fragrance. It's good. Very good.

When I posed for this picture, the art director insisted that I wear it while the pictures were being taken. I thought it was silly, but I said "What the hell? It's their money."

After a while, I realized I like this fragrance a lot. When the photo shoot was over, I walked right over, picked up the bottle, put it in my pocket and said "If you don't mind, I'd like to take this as a souvenir." Then I smiled my best f—— you smile and walked out.

Next time, I'll pay for it.

It's that good.

Today, good-looking straight guys are flocking to the modeling agencies, much less concerned about any homosexual taint that will cleave to them. It's no longer necessary for an ad to plant its tongue firmly in cheek when lavishing erotic attention on the male body—or to pepper the ad with proofs of heterosexuality. It used to be, if an advertisement aimed at straight men dared to show a man fussing over his looks with seemingly romantic plans in mind, there had better be a woman in the picture, making it clear just *whom* the boy was getting pretty for. To sell a muscle-building product to heterosexuals, of course, you had to link it to virility and the ability to attract women on the beach. Today, muscles are openly sold for their looks; Chroma Lean nutritional supplement unabashedly compares the well-sculpted male body to a work of art (and a gay male icon, to boot)—Michelangelo's "David." Many ads display the naked male body without shame or plot excuse, and often exploit rather than resolve the sexual ambiguity that is generated.

Today, too, the athletic, muscular male body that Calvin plastered all over buildings, magazines, and subway stops has become an aesthetic norm, for straights as well as gays. "No pecs, no sex," is how the trendy David Barton gym sells itself: "My motto is not 'Be healthy'; it's 'Look better naked,' " Barton says. The notion has even made its way into that most determinedly heterosexual of contexts, a Rob Reiner film. In *Sleepless in Seattle*, Tom Hanks's character, who hasn't been on a date in fifteen years, asks his friend (played by Rob) what women are looking for nowadays. "Pecs and a cute butt," his friend replied without hesitation. "You can't even turn on the news

nowadays without hearing about how some babe thought some guy's
butt was cute. Who the first woman to say this was I don't know, but
somehow it caught on." Should we tell Rob that it wasn't a woman
who started the craze for men's butts?

rocks and leaners

We "nouvelles voyeuses" thus owe a big measure of thanks to gay
male designers and consumers, and to the aesthetic and erotic over-
lap—not uniform or total, but significant—in what makes our hearts
go thump. But although I've been using the term for convenience, I
don't think it's correct to say that these ads depict men as "sex objects." Actually, I find that whole notion misleading, whether applied
to men or women, because it seems to suggest that what these representations offer is a body that is inert, depersonalized, flat, a mere
thing. In fact, advertisers put a huge amount of time, money, and creativity into figuring out how to create images of beautiful bodies that
are heavy on attitude, style, associations with pleasure, success, happiness. The most compelling images are suffused with "subjectivity"—
they *speak* to us, they seduce us. Unlike other kinds of "objects"
(chairs and tables, for example), they don't let us use them in any way
we like. In fact, they exert considerable power over us—over our psyches, our desires, our self-image.

How do male bodies in the ads speak to us nowadays? In a variety
of ways. Sometimes the message is challenging, aggressive. Many models stare coldly at the viewer, defying the observer to view them in any
way other than how they have chosen to present themselves: as powerful, armored, emotionally impenetrable. "I am a rock," their bodies
(and sometimes their genitals) seem to proclaim. Often, as in the Jackson Browne look-alike ad, the penis is prominent, but *unlike* the penis
in that ad, its presence is martial rather than sensual. Overall, these
ads depict what I would describe as "face-off masculinity," in which
victory goes to the dominant contestant in a game of will against will.
Who can stare the other man down? Who will avert his eyes first?
Whose gaze will be triumphant? Such moments—"facing up," "facing
off," "staring down"—as anthropologist David Gilmore has documented, are a test of macho in many cultures, including our own.

Face-off masculinity

"Don't eyeball me!" barks the sergeant to his cadets in training in *An Officer and a Gentleman*; the authority of the stare is a prize to be won only with full manhood. Before then, it is a mark of insolence—or stupidity, failure to understand the codes of masculine rank. In *Get Shorty*, an unsuspecting film director challenges a mob boss to look him in the eye; in return, he is hurled across the room and has his fingers broken.

"Face-off" ads, except for their innovations in the amount of skin exposed, are pretty traditional—one might even say primal—in their conception of masculinity. Many other species use staring to establish dominance, and not only our close primate relatives. It's how my Jack Russell terrier intimidates my male collie, who weighs over four times as much as the little guy but cowers under the authority of the terrier's macho stare. In the doggie world, size doesn't matter; it's the power of the gaze—which indicates the power to stand one's ground—that counts. My little terrier's dominance, in other words, is based on a convincing acting job—and it's one that is very similar, according to William Pollack, to the kind of performance that young boys in our culture must learn to master. Pollack's studies of boys suggest that a set of rules—which he calls "The Boy Code"—govern their behavior with each other. The first imperative of the code—"Be a sturdy oak"—represents the emotional equivalent of "face-off masculinity": Never reveal weakness. Pretend to be confident even though you may be scared. Act like a rock even when you feel shaky. Dare others to challenge your position.

The face-off is not the only available posture for male bodies in ads today. Another possibility is what I call "the lean"—because these bodies are almost always reclining, leaning against, or propped up against something in the fashion typical of women's bodies. James Dean was probably our first pop-culture "leaner"; he made it stylish for teenagers to slouch. Dean, however, never posed as languidly or was as openly seductive as some of the high-fashion leaners are today. A recent Calvin Klein "Escape" ad depicts a young, sensuous-looking man leaning against a wall, arm raised, dark underarm hair exposed. His eyes seek out the imagined viewer, soberly but flirtatiously. "*Take Me*," the copy reads.

Languid leaners have actually been around for a long time. Statues of sleeping fauns, their bodies draped languorously, exist in classical

take me

art alongside more heroic models of male beauty. I find it interesting, though, that Klein has chosen Mr. Take Me to advertise a perfume called "Escape." Klein's "Eternity" ads usually depict happy, heterosexual couples, often with a child. "Obsession" has always been cutting-edge, sexually ambiguous erotica. This ad, featuring a man offering himself up seductively, invitingly to the observer, promises "escape." From what? *To* what? Men have complained, justly, about the burden of always having to be the sexual initiator, the pursuer, the one of whom sexual "performance" is expected. Perhaps the escape is from these burdens, and toward the freedom to indulge in some of the more receptive pleasures traditionally reserved for women. The pleasures, not of staring someone down but of feeling one's body caressed by another's eyes, of being the one who receives the awaited call rather than the one who must build up the nerve to make the call, the one who doesn't have to hump and pump, but is permitted to lie quietly, engrossed in reverie and sensation.

Some people describe these receptive pleasures as "passive"—which gives them a bad press with men, and is just plain inaccurate too. "Passive" hardly describes what's going on when one person offers himself or herself to another. Inviting, receiving, responding—these are active behaviors too, and rather thrilling ones. It's a macho bias to view the only *real* activity as that which takes, invades, aggresses. It's a bias, however, that's been with us for a long time, in both straight and gay cultures. In many Latin cultures, it's not a disgrace to sleep with other men, so long as one is *activo* (or *machista*)—the penetrator rather than the penetratee. To be a *pasivo*, on the other hand, is to be socially stigmatized. It's that way in prison cultures too—a good indication of the power hierarchies involved. These hierarchies date back to the ancient Greeks, who believed that passivity, receptivity, penetrability were marks of inferior feminine being. The qualities were inherent in women; it was our nature to be passively controlled by our sexual needs. (Unlike us, the Greeks viewed women—not men—as the animalistic ones.) Real Men, who unlike women had the necessary rationality and will, were expected to be judicious in the exercise of their desires. But being judicious and being "active"—deciding when to pursue, whom to pursue, making advances, pleading one's case—went hand in hand.

Allowing oneself to be pursued, flirting, accepting the advances of another, offering one's body—these behaviors were permitted also (but only on a temporary basis) to still-developing, younger men. These young men—not little boys, as is sometimes incorrectly believed—were the true "sex objects" of elite Greek culture. Full-fledged male citizens, on the other hand, were expected to be "active," initiators, the penetrators not the penetratees, masters of their own desires rather than the objects of another's. Plato's *Symposium* is full of speeches on the different sexual behaviors appropriate to adult men with full beards and established professions and glamorous young men still revered more for their beauty than their minds. But even youth could not make it okay for a man to behave *too much* like a woman. The admirable youth was the one who—unlike a woman—was able to remain sexually "cool" and remote, to keep his wits about him. "Letting go" was not seemly.

Where does our culture stand today with respect to these ideas about men's sexuality? Well, to begin with, consider how rarely male actors are shown—on their faces, in their utterances, and not merely in the movements of their bodies—having orgasms. In sex scenes, the moanings and writhings of the female partner have become the conventional cinematic code for heterosexual ecstasy and climax. The male's participation is largely represented by caressing hands, humping buttocks, and—on rare occasions—a facial expression of intense concentration. She's transported to another world; he's the pilot of the ship that takes her there. When men are shown being transported themselves, it's usually been played for comedy (as in Al Pacino's shrieks in *Frankie and Johnny*, Eddie Murphy's moanings in *Boomerang*, Kevin Kline's contortions in *A Fish Called Wanda*), or it's coded to suggest that something is not quite normal about the man—he's sexually enslaved, for example (as with Jeremy Irons in *Damage*). Mostly, men's bodies are presented like action-hero toys—wind them up and watch them perform.

Hollywood—still an overwhelmingly straight-male-dominated industry—is clearly not yet ready to show us a man "passively" giving himself over to another, at least not when the actors in question are our cultural icons. Too feminine. Too suggestive, metaphorically speaking, of penetration by another. But perhaps fashion ads are less uptight? I decided to perform an experiment. I grouped ads that I had collected over recent years into a pile of "rocks" and a pile of "leaners" and found, not surprisingly, that both race and age played a role. African-American models, whether in *Esquire* or *Vibe*, are almost always posed facing-off. And leaners tend to be younger than rocks. Both in gay publications and straight ones, the more languid, come-hither poses in advertisements are of boys and very young men. Once a certain maturity line is crossed, the challenging stares, the "face-off" postures are the norm. What does one learn from these ads? Well, I wouldn't want to claim too much. It used to be that one could tell a lot about gender and race from looking at ads. Racial stereotypes were transparent, the established formulas for representing men and women were pretty clear (sociologist Erving Goffman even called ads "gender advertisements"), and when the conventions were defied it was usually because advertisers sensed (or discovered in their polls) that social shifts had made consumers ready to receive new images. In this "post-modern" age, it's more of a free-for-all, and images are often more re-

A youthful, androgynous "leaner"—appropriately enough, advertising fragrance "for a man or a woman"

active to each other than to social change. It's the viewers' jaded eye, not their social prejudices, that is the prime consideration of every ad campaign, and advertisers are quick to tap into taboos, to defy expectations, simply in order to produce new and arresting images. So it wouldn't surprise me if we soon find languid black men and hairy-chested leaners in the pages of *Gentlemen's Quarterly*.

But I haven't seen any yet. At the very least, the current scene suggests that even in this era of postmodern pastiche racial clichés and gender taboos persist; among them, we don't want grown men to appear too much the "passive" objects of another's sexual gaze, another's desires. We appear, still, to have somewhat different rules for boys and men. As in ancient Greece, boys are permitted to be seductive, playful, to flirt with being "taken." *Men* must still be in command. Leonardo DiCaprio, watch out. Your days may be numbered.

"honey, what do i want to wear?"

Just as fifties masculinity was fought over (metaphorically speaking) by Stanley Kowalski and Stanley Banks, the male fashion scene of the nineties involves a kind of contest for the souls of men too. Calvin Klein, Versace, Gucci, Abercrombie & Fitch have not only brought naked bottoms and bulging briefs onto the commercial scene, they present underwear, jeans, shirts, and suits as items for enhancing a man's appearance and sexual appeal. They suggest it's fine for a man to care about how he looks and to cultivate an openly erotic style. In response, aggressively heterosexual Dockers and Haggar ads compete— for the buying dollar of men, but in the process for their gender consciousness too—by stressing the no-nonsense utility of khakis. Consider the Haggar casuals advertisement on the next page, and what it says about how "real men" should feel about their clothes:

"I'm damn well gonna wear what I want. . . . Honey, what do I want?"

Looked at in one light, the man in the advertisement is being made fun of, as a self-deceived blusterer who asserts his independence "like a man" and in the next breath reveals that he is actually a helpless little boy who needs his mommy to pick out his clothes for him. But fashion incompetence is a species of helplessness that many men feel

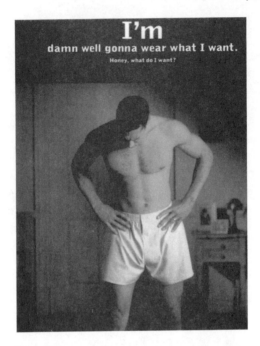

quite comfortable with, even proud of. Recognizing this, Haggar and Dockers are among those manufacturers who have put a great deal of effort into marketing "nonfashion-guy fashion" to a niche of straight men—working-class and yuppie—who, they presume, would be scared off by even a whiff of "feminine" clothes-consciousness. Here's another one from Haggar's:

"*In the* female *the ability to match colors comes at an early age. In the* male *it comes when he marries a female.*"

The juxtaposition of inept male/fashion-conscious female, which with one stroke establishes the masculinity *and* the heterosexuality of the depicted man, is a staple of virtually every Haggar ad. In a Haggar television spot with voice-over by John Goodman (Roseanne's beefy former television husband), a man wakes up, sleepily pulls on a pair of khakis, and goes outside to get the paper:

"*I am not what I wear. I'm not a pair of pants, or a shirt.*" (He then walks by his wife, handing her the front section of the paper.) "*I'm not in touch with my inner child. I don't read poetry, and I'm not politically correct.*" (He goes down a hall, and his kid snatches the comics from him.) *"I'm just a guy, and I don't have time to think about what I wear, because I've got a lot of important guy things to do."* (Left with only the sports section of the paper, he heads for the bathroom.) *"One-hundred-per-cent-cotton-wrinkle-free khaki pants that don't require a lot of thought. Haggar. Stuff you can wear."*

Yes, it's a bit of a parody, but that only allows Haggar to double its point that real guys have better things to do than think about what they are going to wear or how they appear to others. The guy who would be so worried about his image that he couldn't poke fun at himself wouldn't be a real guy at all. Real guys don't take themselves so seriously! That's for wimps who favor poetry, self-help psychology, and bleeding-heart politics. That's for girls, and for the men who are pussy-whipped by them.

In Haggar's world, real guys don't choose clothing that will enhance the appearance of their bodies or display a sense of style; real guys just put on some "stuff" to wear because they have to, it's socially required. The less decorative, the better. "We would never do anything with our pants that would frighten anyone away," says Dockers designer Gareth Morris as reported in a 1997 piece in *The New Yorker*. "We'd never do too many belt loops, or an unusual base cloth . . . [or] zips or a lot of pocket flaps and details on the back." Pocket flaps, the ultimate signifier of suspect sexuality! In such ads, male naiveté about the sexual potency of clothes, as agency maven David Altschiller claims, is critical. "In women's advertising," he points out, "self-confidence is sexy. But if a man is self-confident—if he knows he is attractive and is beautifully dressed—then he's not a man anymore. He's a fop. He's effeminate." In Dockers' "Nice Pants" television ads, for example, it's crucial that the guy not *know* his pants are "nice" until a gorgeous woman points it out to him.

It's no accident that the pants are described via the low-key understatement "nice" (rather than "great," for example, which would suggest that the guy was actually *trying* to look good). For the real man (according to Dockers), the mirror is a tool, not a captivating pool; if he could, he'd look the other way while he shaves. Many other advertisers capitalize on such notions, encouraging men to take care of their looks, but reassuring them that it's for utilitarian or instrumental purposes. Cosmetic surgeons emphasize the corporate advantage that a face-lift or tummy tuck will give the aging executive: "A youthful look," as one says, "gives the appearance of a more dynamic, charging individual who will go out and get the business." Male grooming products too are often marketed by way of "action hero" euphemisms which obscure their relation to feminine versions of the same product (a male girdle marketed by BodySlimmers is called the Double Agent Boxer) and the fact that their function is to enhance a man's appearance: hair spray as "hair control," exfoliating liquid as "scruffing lotion," astringents as "scrubs," moisturizers and fragrances as "after" or "pre" accompaniments to that most manly of rituals, the shave. They often have names like Safari and Chaps and Lab Series, and come in containers shaped like spaceships and other forms a girl could have some fun with.

The notions about gender that are maintained in this marketing run deeper than a refusal to use the word "perfume" for products designed to make men smell good. In the late seventies, coincident with the development of feminist consciousness about these matters, art historian John Berger discovered what he argued were a set of implicit cultural paradigms of masculinity and femininity, crystallized in a visual "rule" of both classical painting and commercial advertisements: *"men act and women appear."* Here's a contemporary illustration:

The man in the *Nautica* ad on the facing page, rigging his sail, seems oblivious to his appearance; he's too busy checking the prevailing winds. The woman, in contrast, seems well aware and well pleased that her legs have caught the attention of the men gaping at her. A woman's *appearance*, Berger argued, has been socially determined to be "of crucial importance for what is normally thought of as the success of her life." Even walking on a city street, headed for their high-powered executive jobs, women exist to be seen, and they *know* it—a notion communicated by the constant tropes of female narcissism: women shown preening, looking in mirrors, stroking their own bodies, exhibiting themselves for an assumed spectator, asking to be admired for their beauty.

With depictions of men, it's just the opposite. "A man's presence," Berger wrote, "is dependent upon the promise of power which he embodies . . . what he is capable of doing to you or for you." Thus, the classic formula for representing men is always to show them in action, immersed in whatever they are doing, seemingly unaware of anyone who might be looking at them. They never fondle their own bodies

Men act and women appear

narcissistically, display themselves purely as "sights," or gaze at them-selves in the mirror. In everything from war paintings to jeans and cologne ads, men have been portrayed as utterly oblivious to their beauty (or lack of it), intent only on getting the job done—raising the flag, baling hay, lassoing a steer, busting up concrete. The ability to move heavy things around, tame wild creatures—that's manly busi-ness. Fretting about your love handles, your dry skin, your sagging eyelids? That's for girls.

Women in ads and movies thus require no plot excuse to show off their various body parts in ads, proudly, shyly, or seductively; it's the "business" of *all* of us to be beautiful—whether we are actresses, politicians, homemakers, teachers, or rock stars. This has changed very little since Berger came up with his formula. When *Time* magazine did a story on the new dominance of female stars in the rock world, its cover featured singing star Jewel, not performing, but in a dewy close-up, lips moist and soft eyes smiling from behind curled lashes. This formidable new "force" in the rock world might as well have been modeling Maybelline. True, a beautiful woman today may be depicted puffing away on a cigar, getting "in touch with her masculine side." But in expression she's still a seductress, gazing through long-lashed lids into the eyes of an imagined viewer. "Do you like what you see?" the expressions of the models seem to ask.

Men, according to Berger's formula, must never seem as though they are asking this question, and may display their beauty only if it is an unavoidable side effect of other "business." Thus, a lot of the glistening, naked male chests in the movies of the fifties and sixties were on the bodies of warriors, prisoners, slaves, and prizefighters. No one could claim there was vanity in such nakedness. (No time for preening while nailing spikes on a chain gang or rowing in a slave galley.) So a strong dose of male skin could be sneaked into a movie without disturbing the gender rules. The physical presence of an actor like Richard Gere, who emanates consciousness of his body as the erotic focus of the gaze and invites it, has always annoyed and disconcerted critics. The pomposity of Charlton Heston, on the other hand, his naked (and actually rather gorgeous) chest puffed up in numerous biblical epics, goes unnoticed, because he's doing it all in a builder-of-the-universe rather than narcissus-in-the-mirror mode.

Saturday Night Fever (1977) deserves mention here, for openly breaking with this convention. Tony Manero (John Travolta), a disco-dancing dandy who knows how to use his walk, was a man who *really* needed a course in masculinity-according-to-Haggar. He blows all his wages on fancy shirts and shoes. On Saturday night, he prepares his body meticulously, shaving, deodorizing, blow-drying, choosing just the right combination of gold chains and amulets, torso-clinging pants, shiny platforms. Eating dinner with his family, he swathes himself in a sheet like a baby to protect his new floral shirt; when his father boxes his ear roughly, his only thought is for his pompadour: "Just watch the hair! I work on my hair a long time and you hit it. He hits the hair!" Manero spends much of his time in front of the mirror, getting himself pretty, posing, anticipating the impression he's going to make when he enters the disco or struts down the street.

Never before *Saturday Night Fever* had a heterosexual male movie hero spent so much time on his toilette. (Even Cary Grant's glamorous looks were never shown as requiring any conscious effort or attention; in *The Awful Truth* he sits under a tanning lamp—but that's to fake a trip to Florida.) Although this was the polyester seventies, and men like Sonny Bono dressed like Tony on television, Bono was very careful (as the Beatles were too) to treat his flamboyant ruffles as showbiz costumes, while Cher proudly strutted her feathers and finery as a second skin for her body and sexuality. Tony, like Cher, chooses his clothes to highlight his sinuous form.

Manero was, in many ways, the cinema equivalent (reassuringly straight and working-class) of the revolution that Calvin Klein was making in more sexually ambiguous form in the fashion world. As a dancer, Tony is unembarrassed—and the camera isn't embarrassed either—to make his hips, groin, and buttocks the mesmerizing center of attention. Travolta was also the first actor to appear on-screen in form-fitting (if discreetly black) briefs. One scene finds him asleep in his underwear, blanket between his legs, hip jutting upward; the camera moves slowly down the length of his body, watches as Tony rouses, sits up, pulls the blanket from between his legs, and puts his hand in his briefs to adjust his penis. (The script originally had called for Travolta to appear naked in a later scene; he balked, suggesting the early morning scene as a compromise.) We then follow him to the mirror (where he compares himself admiringly with a poster of Al Pacino) and into the hall, where he flexes teasingly for his shocked grandmother. This was new stuff, and some people were a bit taken aback by such open male vanity and exhibitionism. (Pauline Kael, for one, seemed to need to convince herself of Tony's sexual orientation. "It's a straight heterosexual film," she wrote, "but with a feeling for the sexiness of young boys who are bursting their britches with energy and desire.")

True, there is the suggestion, in the film, that Tony may grow out of his narcissism once he leaves Brooklyn and the gold chain crowd. Hollywood, of course, had shown men preening, decorating, and oiling themselves before—pimps and homosexuals, usually, but also various unassimilated natives (blacks, Puerto Ricans, Italians) depicted as living more fully in their bodies, with a taste for flashy clothes that marks them as déclassé. Manero fits those stereotypes—but only up to a point. He may have awful taste in jewelry, but he also has boyish charm and "native" intelligence. Unlike his friends—a pathetic trio of racist, homophobic, sexist homeboys—Tony has integrity. He is enraged when, at the "2001" dance contest, racism and favoritism land him first prize over a Puerto Rican couple. He's also the only one of his friends who doesn't taunt a gay couple as they pass on the street. The movie may poke affectionate fun at him, but it also admires him. A hero-narcissus—a very new image for postwar Hollywood.

Of course, most men, gold chains or not, straight or gay, *do* care how they "appear." The gender differences described in Berger's formula and embedded in the Dockers and Haggar advertisements are "fictional," a distillation of certain *ideas* about men and women, not an empirical generalization about their actual behavior. This doesn't mean, however, that they have no impact on "real life." Far from it. As embodied in attractive and sometimes highly manipulative images, "men act and women appear" functions as a visual instruction. Women are supposed to care very much about fashion, "vanity," looking good, and may be seen as unfeminine, man-hating, or lesbian if they don't. The reverse goes for men. The man who cares about his looks the way a woman does, self-esteem on the line, ready to be shattered at the slightest insult or weight gain, is unmanly, sexually suspect.

So the next time you see a Dockers or Haggar ad, think of it not only as an advertisement for khakis but also as an advertisement for a certain notion of what it means to be a man. The ad execs know that's what's going on, they're open about not wanting to frighten men off with touches of feminine decorativeness. What they are less open about is the fact that such ads don't just cater to male phobias about fashion but also perpetuate them. They have to. Nowadays, the Dockers man is competing against other models of masculinity, laughing at him from both the pages of history and from what was previously the "margin" of contemporary culture. Can you imagine Cary Grant, Rupert Everett, or Michael Jordan as the fashion-incompetent man in a Dockers ad? The stylish man, who began to make a new claim on popular cultural representations with the greater visibility of black and gay men—the men consumer culture once ignored—was chiseling cracks in the rule that "men act and women appear" even as Berger was formulating it.

male decorativeness in cultural perspective

Not all heterosexual men are as uptight about the pocket flaps on their pants as the Haggar executive would have us believe. Several weeks after the piece on khakis appeared in *The New Yorker*, a reader wrote in protesting that the idea "that men don't want to look like they're trying to be fashionable or sexy" was rather culture-bound. Maybe, this reader acknowledged, it applies to American, English, and Japanese men. "But are we really to believe that French, Italian, and Spanish men share this concern? And, when we expand the category 'male' beyond human beings, biologists have shown that the demonstration of male splendor is a key element in the vertebrate mating game. Are American males just an anomalous species?"

The letter reminds us that there are dangers in drawing broad conclusions on the basis of only those worlds with which one is familiar. And it's not just different international attitudes toward men and fashion that cast doubt on the universal applicability of the Dockers/Haggar view of masculinity. To look at the variables of race, class, and history is to produce a picture of male attitudes toward fashionable display that is far from consistently phobic.

First of all, for most of human history, there haven't been radically different "masculine" and "feminine" attitudes toward beauty and decorativeness. On farms, frontiers, and feudal estates, women were needed to work alongside men and beauty was hardly a priority for either. Among aristocrats, it was most important to maintain class privilege (rather than gender difference), and standards of elegance for both sexes (as Anne Hollander's fascinating *Sex and Suits* documents) were largely the same: elaborate headwear, cosmetics, nonutilitarian adornments, and accessories. Attention to beauty was associated not with femininity but with a life that was both privileged and governed by exacting standards. The constrictions, precarious adornments, elaborate fastenings reminded the elite that they were highly civilized beings, not simple peasant "animals." At the same time, decorativeness was a mode of royal and aristocratic competition, as households and courts would try to out-glam each other with jewels and furs. Hollander describes a sixteenth-century summit meeting between Francis I and Henry VIII, in which everyone wore "silver covered with diamonds, except when they were in cloth of gold and covered with rubies. Everything was lined with ermine and everything was 20 yards long, and there were plumes on everybody." Everybody—male or female—had to be as gorgeous as possible. It was a mode of power competition.

Until roughly the fourteenth century, men and women didn't even

dress very differently. (Think of the Greeks and Romans and their uni-
sex robes and togas.) Clear differences started to emerge only in the
late Middle Ages and early Renaissance: women's breasts began to be
exposed and emphasized in tight bodices, while their legs were covered
with long skirts. Men's legs—and sometimes their genitals as well—
were "fully articulated" and visible through pantaloons (what we call
"tights"), with body armor covering the chest. While to our sensibilities, the shapely legs and genitals of men in tights (unless required by a
ballet or historical drama) are either to be laughed at or drooled over,
Hollander argues that in the Renaissance, to outline the male body
was to make it more "real" and "natural," less a template for sexual
fantasy (as women's bodies were becoming). This trend continued,
with men's clothing getting progressively more unrestrictive, tailored,
simple and women's more stiff, tightly fitted, decorative. Still, into the
seventeenth century, fashionable gentleman continued to wear lace and
silk, and to don powder and wigs before appearing in public. Hollander regards the nineteenth century as a "great divide," after which not
only the styles of men's and women's clothing (trousers for men, increasingly romantic froufrou for women) would become radically different, but ideas about them as well. Men's clothing must now be
"honest, comfortable, and utilitarian," while women's begins to develop a reputation for being "frivolous" and "deceptive." The script
for "men act and women appear" was being written—right onto male
and female clothing.

Looking beyond fashion to the social world (something Hollander
refuses to do, but I'll venture), it's hard not to speculate that these
changes anticipate the emergence of the middle class and the nineteenth-century development of distinctively separate spheres for men
and women within it. In the industrial era, men's sphere—increasingly
the world of manufacturing, buying, selling, power brokering—was
performance-oriented, and demanded "no nonsense." Women, for
their part, were expected not only to provide a comfortable, well-
ordered home for men to return to but to offer beauty, fantasy, and
charm for a man to "escape" to and restore himself with after the grim
grind of the working day. As this division of labor developed, strong
dualistic notions about "masculinity" and "femininity" began to
emerge, with sanctions against the man or woman who dared to cross
over to the side of the divide where they did not belong "by nature."

By the end of the nineteenth century, older notions of manliness premised on altruism, self-restraint, and moral integrity—qualities that women could have too—began to be understood as vaguely "feminine." Writers and politicians (like Teddy Roosevelt) began to complain loudly about the emasculating effects of civilization and the excessive role played by women teachers in stifling the development of male nature. New words like "pussyfoot" and "stuffed shirt"—and, most deadly, "sissy"—came into parlance, and the "homosexual" came to be classified as a perverse personality type which the normal, heterosexual male had to prove himself distinct from. (Before, men's relations with each other had been considerably more fluid, and even the heterosexual male was allowed a certain degree of physical intimacy and emotional connection—indeed, "heterosexuality" as such was a notion that hardly made sense at the time.) A new vogue for bodybuilding emerged. "Women pity weakly men," O. S. Fowler warned, but they love and admire "right hearty feeders, not dainty; sprightly, not tottering; more muscular than exquisite, and more powerful than effeminate, in mind and body." To be "exquisite," to be decorative, to be on display, was now fully woman's business, and the man who crossed that line was a "fop."

From that time on, male "vanity" went into hiding, and when cosmetic products for men began to be marketed (for men *did* use them, albeit in secret), they had to justify themselves, as Kathy Piess documents, through the manly rhetoric of efficiency, rugged individualism, competitive advantage, autonomy. While Pompeian cream promises to "beautify and youthify" women, the same product for men will help them "win success" and "make promotion easier" on the job. Even that most manly of rituals (from our perspective), shaving, required special rhetoric when home shaving was first introduced early in the twentieth century. "The Gillette is typical of the American spirit," claimed a 1910 ad. "Its use starts habits of energy—of initiative. And men who *do* for themselves are men who *think* for themselves." Curley's Easy-Shaving Safety Razor claimed that "the first Roman to shave every day was no fop, but Scipio, conqueror of Africa." When it came to products used also by women—like scents and creams—manufacturers went out of their way to reassure prospective customers of their no-nonsense "difference," through action names (Brisk, Dash, Vim, Keen, Zest) and other means. When Florian, a line of men's toiletries, was introduced in 1929, its creator, Carl Weeks, advised druggists to locate the products near cigar (again!) counters, using displays featuring manly accouterments like boxing gloves, pipes, footballs. This, he argued, "will put over the idea that the mascu-*line* is all *stag*. It's for he-men with no women welcome nohow."

This isn't to say that from the turn of the nineteenth century on, the drive to separate "masculine" and "feminine" attitudes toward self-beautification pushed forward relentlessly. For one thing, culture is never of one piece; it has its dominant images, but also its marginal, recessive, and countercultural images. For another, the history of gender ideology didn't end with the nineteenth century, as dramatic as its changes were. A century of mutations and permutations followed, as demanded by social, economic, and political conditions. Older ideals lingered too and were revived when needed. The Depression, for example, brought a love affair with (a fantasy of) aristocratic "class" to popular culture, and a world of Hollywood representations—as we've seen—in which sexual difference was largely irrelevant, the heroes and heroines of screwball comedy a matched set of glamorously attired cut-ups. In these films, the appeal of actors like Cary Grant, Fred Astaire, and William Powell was largely premised not on assertions of masculine performance but on their elegance, wit, and charm. Their maleness wasn't thrown into question by the cut of their suits. Rather, being fashionable signified that they led an enviable life of pleasure and play. Such associations still persist today. Fashion advertisements for Ralph Lauren, Valentino, Hugo Boss, and many others are crafted to appeal to the class consciousness of consumers; in that universe, one can never be too beautiful or too vain, whatever one's sex.

In the screwball comedies, it didn't matter whether you were a man or a woman, everyone's clothes sparkled and shone. Following the lead of the movies, many advertisements of the thirties promoted a kind of androgynous elegance. But others tried to have their cake and eat it too, as in a 1934 ad for Fougère Royale aftershave, which depicts a group of tony men in tuxedos, hair slicked back, one even wearing a pince-nez, but with the caption "Let's *not* join the ladies!" We may be glamorous, even foppish—but *puh-lease! Ladies* we're not! I should note, too, that while the symbols of "class" can function to highlight equality between men and women, they can also be used to emphasize man's superiority over women—as in a contemporary Cutty Sark ad in which a glamorously attired woman relaxes, dreamily stroking a dog, while the tuxedo-clad men standing around her engage in serious conversation (about stocks, I imagine); these guys don't need to go off into the drawing room in order to escape the ladies; they can keep one around for a bit of decorativeness and sensual pleasure while she remains in her own, more languorous world within their own.

During World War II, movies and magazines continued to celebrate independent, adventurous women, to whom men were drawn "as much for their spirit and character as for their looks."* But when the fighting men returned, the old Victorian division of labor was revived with a new commercial avidity, and the world became one in which "men act" (read: *work*) and "women appear" (read: *decorate*—both themselves and their houses)—with a vengeance. Would Barbie get on a horse without the proper accessories? Would the Marlboro Man carry a mirror with him on the trail? By the late fifties and early sixties, the sexy, wisecracking, independent-minded heroine had morphed into a perky little ingenue. Popular actresses Annette Funicello, Connie Stevens, and Sandra Dee were living Barbie dolls, their femininity blatantly advertised on their shirtwaisted bodies. They had perfectly tended bouffant hairdos (which I achieved for myself by sleeping on the cardboard cylinders from toilet tissue rolls) and wore high heels even when washing dishes (I drew the line at that). And what about the dashing, cosmopolitan male figure in fashionable clothes? He now was usually played as a sissy or a heel—as for example Lester (Bob Evans), the slick playboy of *The Best of Everything*, who seduces gullible April (Diane Baker) with his big-city charm, then behaves like a cad when she get pregnant.

There have always been ways to market male clothes consciousness, however. Emphasizing neatness is one. Our very own Ronnie Reagan (when he was still a B-movie star) advertised Van Heusen shirts as "the neatest Christmas gift of all" because they "won't wrinkle . . . ever!!"

---

*Not that women's beauty was dispensable. Concern for her looks symbolized that although she worked as hard as a man, a woman's mind was still on the *real* men who were fighting for her freedom. (An ad for Tangee lipstick describes "a woman's lipstick [as] an instrument of personal morale that helps her to conceal heartbreak or sorrow; gives her self-confidence when it's badly needed . . . It symbolizes one of the reasons why we are fighting . . . the precious right of women to be feminine and lovely—under any circumstances.") The woman of this period was a creature of both "appearance" *and* "action"—a kind of forerunner to today's superwoman.

Joining elegance with violence is another. James Bond could get away with wearing beautiful suits because he was ruthless when it came to killing and bedding. (A men's cologne, called 007, was advertised in the sixties with clips from *Thunderball*, the voice-over recommending: "When you use 007, be kind" because "it's loaded" and "licensed to kill . . . women.") The elegant male who is capable of killing is like the highly efficient secretary who takes off her glasses to reveal a passionate, gorgeous babe underneath: a species of tantalizing, sexy disguise.

When elegance marks one man's superior class status over another it gives him a competitive edge (as was the dominant function of elegance before the eighteenth century) rather than turning him into a fop. "We have our caste marks, too" ran a 1928 ad for Aqua Velva, which featured a clean-shaven, top-hatted young man, alongside a turbaned, bejeweled, elite Indian man. This ad, however, proved to be problematic, as Kathy Piess points out. American men didn't like being compared with dark-skinned foreigners, even aristocratic ones. The more dominant tradition—among Europeans as well as Americans— has been to portray an order in which the clean, well-shaven white man is being served or serviced by the dark ones, as in a 1935 American ad for Arrow Shirts in which the black maid is so fashion-clueless that she doesn't even know what a manufacturer's label is, or in a German ad for shaving soap depicting the "appropriate" relation between the master race and the Others.

Such codes were clearly being poked fun at—how successfully I'm not sure—when a 1995 Arid Extra-Dry commercial depicted African-American pro basketball player Charles Barkley dressed up as a nineteenth-century British colonial, declaring that anything less than Arid "would be uncivilized." The commercial, however, is not just (arguably) a poke at the racist equation of civilization and whiteness. It's also, more subtly, a playful assertion of some distinctive African-American attitudes toward male display. "Primordial perspiration," Barkley says in the commercial, "shouldn't mess with your style." And "style" is a concept whose history and cultural meanings are very different for blacks and whites in this country. Among many young African-American men, appearing in high style, "cleaned up" and festooned with sparkling jewelry, is not a sign of effeminacy, but potency and social standing. Consider the following description, from journalist Playthell Benjamin's 1994 memoir, *Lush Life* (while you're reading

1935

it, you might also recall Anne Hollander's description of Henry VIII's summit meeting):

> [Fast Black] was dressed in a pair of white pants, white buck shoes, and a long-sleeve white silk shirt—which was open to his navel and revealed a 24-karat gold chain from which hung a gold medallion set with precious stones: diamonds, rubies, and emeralds. His massively muscled body was strikingly displayed in a white see-through silk shirt, and the trousers strained to contain his linebacker thighs. His eyes were bloodshot and his skin was tight against his face, giving it the look of an ebony mask. He struck me right off as a real dangerous muthafucka; mean enough to kill a rock.

A "real dangerous muthafucka" in a white see-through silk shirt? For the white boys to whom the Dockers and Haggar ads are largely addressed, see-through silk is for girls, and showing off one's body—particularly with sensuous fabrics—is a "fag" thing. Thus, while a Haggar ad may play up the sensual appeal of soft fabrics—*"These clothes are very soft and they'll never wrinkle"*—it makes sure to include a parenthetical (and sexist) reference to a dreamed-of wife: *"Too bad you can't marry them."* But sartorial sensuality and decorativeness, as I've learned, do not necessarily mean "femininity" for African-American men.

When I first saw the Charles Barkley commercial, the word "style" slipped by me unnoticed, because I knew very little about the history of African-American aesthetics. An early paper of mine dealing with Berger's equation was utterly oblivious to racial differences that might confound the formula "men act and women appear." Luckily, an African-American male colleague of mine gently straightened me out, urging me to think about Mike Tyson's gold front tooth as something other than willful masculine defiance of the tyranny of appearance. Unfortunately, at that time not much of a systematic nature had been written about African-American aesthetics; I had to find illuminating nuggets here and there. Then, just this year, Shane White and Graham White's *Stylin'* appeared. It's a fascinating account of how the distinctive legacy of African aesthetics was maintained and creatively, sometimes challengingly, incorporated into the fashion practices of American blacks, providing a vibrant (and frequently subversive) way for blacks to "write themselves into the American story."

Under slavery, white ownership of blacks was asserted in the most concrete, humiliating way around the display of the body on the auction block. Slaves were often stripped naked and instructed to show their teeth like horses being examined for purchase. Women might have their hair cut off. Everyone's skin would be polished to shine, as apples are polished in grocery stores today. As a former slave described it:

"The first thing they had to do was wash up and clean up real good and take a fat greasy meat skin and run over their hands, face and also their feet, or in other words, every place that showed about their body so that they would look real fat and shiny. Then they would trot them out before their would-be buyers and let them look over us real good, just like you would a bunch of fat cows that you were going to sell on the market and try to get all you could for them."

It makes perfect sense that with the body so intimately and degradingly under the control of the slave owner, opportunities to "take back" one's own body and assert one's own cultural meanings with it would have a special significance. On Sundays, slaves would dress up for church in the most colorful, vibrant clothes they could put together—a temporary escape from and an active repudiation of the subservience their bodies were forced into during the week. Their outfits, to white eyes, seemed "clashing" and mismatched. But putting together unusual combinations of color, texture, and pattern was an essential ingredient of West African textile traditions, handed down and adapted by African-American women. Color and shape "coordination"—the tyranny of European American fashion until pretty recently—were not the ruling principles of style. "Visual aliveness," *Stylin'* reports, was. The visual aliveness of the slaves' Sunday best, so jangling to white sensibilities, was thus the child both of necessity—they were forced to construct their outfits through a process of bricollage, putting them together from whatever items of clothing were available—and aesthetic tradition.

From the start, whites perceived there was something insubordinate going on when blacks dressed up—and they were not entirely wrong. "Slaves were only too keen to display, even to flaunt, their finery both to slaves and to whites"; the Sunday procession was, as I've noted, a time to reclaim the body as one's own. But at the same time, blacks were not just "flaunting," but preserving and improvising on vibrant African elements of style whose "flashiness" and "insolence" were largely in the eye of the white beholder, used to a very different aesthetic. The cultural resistance going on here was therefore much deeper than offended whites (and probably most blacks too) realized at the time. It wasn't simply a matter of refusal to behave like Stepin Fetchit, with head lowered and eyes down. A new culture of unpredictable, playfully decorative, visually bold fashion was being created—and it would ultimately (although not for some time) transform the world of mainstream fashion as much as Klein's deliberately erotic underwear and jeans.

After "emancipation," funeral marches and celebratory promenades were a regular feature of black city life, in which marchers, male and female alike, were "emblazoned in colorful, expensive clothes," the men in "flashy sports outfits: fancy expensive silk shirts, new pants, hats, ties, socks," "yellow trousers and yellow silk shirts," and "bedecked with silk-and-satin-ribboned streamers, badges." Apart from formal processions, streets like Memphis's Beale Street and New Orleans's Decatur Street were ongoing informal sites for "strolling" and display. The most dazzlingly dressed men, often jazz musicians, were known as "sports." As "Jelly Roll" Morton describes it, each "sport" had to have a Sunday suit, with coat and pants that did not match, and crisply pressed trousers as tight as sausage skins. Suspenders were essential and had to be "very loud," with one strap left provocatively "hanging down." These guys knew how to "use their walk" too. The sport would walk down the street in a "very mosey" style: "Your hands is at your sides with your index fingers stuck out and you kind of struts with it." Morton—by all accounts a particularly flashy sport—had gold on his teeth and a diamond in one of them. "Those days," he recalled, "I thought I would die unless I had a hat with the emblem Stetson in it and some Edwin Clapp shoes." Shades of Tony Manero. Or King Henry VIII.

In fact, the flashiest African-American male styles have partaken both of the African legacy and European notions of "class." Although the origin of the zoot suit—broad shoulders, long coats, ballooning, peg-legged trousers, usually worn with a wide-brimmed hat—is debated, one widely believed account says it was based on a style of suit worn by the Duke of Windsor. Another claims Rhett Butler in *Gone With the Wind* was the inspiration for the zoot suit (if so, it is a "deep irony," as the authors of *Stylin'* comment). But whatever its origins, the zoot suit, worn during the forties when cloth conservation orders ruled the use of that much fabric illegal, was a highly visible and dramatic statement in *disunity* and defiance of "American Democracy," a refusal to accede to the requirements of patriotism. Even more so than the slave's Sunday promenade, the zoot-suiter used "style" aggressively to assert opposition to the culture that had made him marginal to begin with—without his assent.

The use of high style for conspicuous display or defiance is still a big part of male street culture, as sociologist Richard Majors notes: "Whether it's your car, your clothes, your young body, your new hairdo, your jewelry, you style it. The word 'style' in [African-American] vernacular usage means to show off what you've got. And for teenagers with little money and few actual possessions, showing off what you do have takes on increased importance. As one youth puts it, 'It's identity. It's a big ego trip.' "

What's changed since Majors wrote these words in the early nineties is the increasing commercial popularity of hip-hop music and culture, which has turned the rebellious stylings of street youth into an empire of

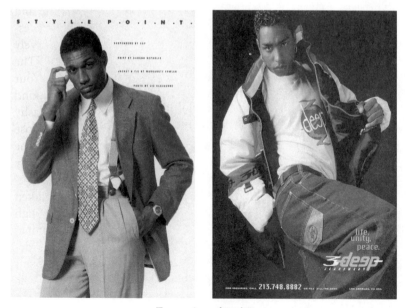

Two versions of "style"

images and products, often promoted (and sometimes designed) by big-name stars. With postmodern sensibilities (grab what you like) ruling the fashion world, moreover, what once were signature elements of black street style have been incorporated—as gay styles have also been incorporated—in the fashions of other worlds, both "high" (designer clothing) and "low" (white high school boys with their pants slung low, trying to look so cool).

Despite the aggressive visibility of hip-hop culture, "showing off what you've got" has not been the only influential definition of style among African-Americans. In the late nineteenth and early twentieth century, several etiquette books were published, written by middle-class blacks, promoting a very different fashion ideal. The *National Capital Code of Etiquette*, published in 1889, warned young men to "avoid colors that do not blend with the remainder of your wearing apparel, and above all things shun the so-called 'loud' ties with colors that fairly shriek unto Heaven . . ." The young black men should also avoid "bright reds, yellows and light greens as you would the plague" and never, ever strut or swagger. Hortense Powdermaker, who studied black life in Indianola, Mississippi, in the late 1930s, noted that better-off African-Americans "deliberately avoided bright colors" and were offended when clerks, on the basis of "the Negroe's reputation for wearing gaudy clothes," assumed they wanted something "loud." Those who advocated a less ostentatious style were dismayed by the lower-class practice of adorning healthy front teeth with gold, while leaving bad back teeth unattended.

A recent *Essence* list of fashion "do's and don'ts" emphasizes this deliberately understated—and in today's world, "professional"—conception of black male style. "Yes" to well-groomed hands, well-fitting suit and a "definite sense of self." "No" to "glossy polished nails," "cologne that arrives before he does," "Mr. T jewelry (the T stands for tacky)," and "saggy jeans on anyone old enough to remember when 'Killing Me Softly' was *first* released." Even in their most muted variations, African-American styles have done a great deal to add color, playfulness, and unexpected, sexy little fillips to "tasteful," professional male clothing: whimsical ties, internationally inspired shirts and sweaters, and, in general, permission to be slightly dramatic, flirtatious, and ironic with one's clothes. The rule of always matching patterns, too, no longer holds in the world of high fashion, the result of a collaboration (not necessarily conscious, of course) between postmodern sensibilities and the slave legacy of bricolage.

Superstar Michael Jordan (his masculine credentials impeccable, his reputation as a family man solidly established over the years), a very effective spokesperson for style, has done a great deal to make fashionableness, even "feminine" decorativeness, congruous with masculinity. This year, he was named *GQ*'s "Most Stylish Man." "How stylish is Michael Jordan?" *GQ* asks. "Answer: So stylish he can get away with wearing five rings!" Of course, the fact that Jordan can "get away" with wearing five rings reveals *GQ*'s cultural biases. For the magazine, Jordan's stylishness resides in the "drape of his suits, in the plain gold hoop in his left ear, in the tempered, toned-down body language of his late career." For *GQ*, subtlety equals style. For Jordan too. But of course that plain gold hoop would not have been viewed as so tastefully subtle had Jordan not made it an acceptable item of male decorativeness.

Jordan, God bless him, is also unabashed in admitting that he shops more than his wife, and that he gets his inspiration from women's magazines. The night before he goes on the road, he tries on every outfit he's going to wear. He describes himself as a "petite-type person" who tries to hide this with oversize clothes and fabrics that drape. When questioned about the contradiction between the "manliness" of sports and his "feminine" love of fashion, Jordan replies that "that's the fun part—I can get away from the stigma of being an athlete." Saved by fashion from the "stigma" of being a sweaty brute—that's something, probably, that only an African-American man can fully appreciate. The fact that it's being an athlete and not "femininity" that's the "stigma" to be avoided by Jordan—that's something a woman's got to love.

The ultimate affront to Dockers masculinity, however, is undoubtedly the Rockport ad on the following page, with drag superstar RuPaul in a beautifully tailored suit. His feet and his stare are planted—vitually identically to Michael Jordan's posture in the feature I've just discussed—in that unmistakable (and here, ironic) grammar of face-off ad masculinity. "I'm comfortable being a MAN," declares RuPaul. "I'm comfortable being a woman too," of course, is the unwritten subtext. Man, woman, what's the difference so long as one is "uncompromising" about style?

man of style

makes the statement: It's clean, classy.

Sports is considered manly and fashion is thought of as feminine. How do you reconcile the two? That's the fun part—I can get away from the stigma of being an athlete. A lot of guys in the pros take that some jol fashion] rery seriously. We're not sweaty all the time, always wearing sweat suits and gym shoes. That's something I learned from Dr. J a long time ago. He's such a classy individual.

How do you assemble your wardrobe? The night before I go on the road, I try on every outfit I'm going to wear: the watch with the suit with the tie, with the socks with the shoes. Then I pack them up that way.

Are suits a sign of aggression? Sure. They let people know I'm serious. My father always said, "Sometimes people's personalities can be defined by the way they're dressed—or the way they smell." And I take that to heart.

Ever give [cross-dressing teammate] Dennis Rodman any pointers? He has his own style [laughs]. I have yet to see something he wears that I would consider wearing. Dennis and I are on totally different pages.... I wouldn't wear stockings.

What do you wear to go out with your wife? I don't wear jeans outside the house. Very few people see me in a suit, they see me in shorts. That's my activity in public. So to see me in a suit, it's different but it's me. It's a part of my personality. I want people to know.

You have just launched your cologne, with socks and deodorant to follow. Why? I have always been intrigued *(continued on page 56)*

(continued on page 56)

THE SUIT
"I'm a suit guy. I have anywhere from $60 to $60 [this one's from the shop Bordi in Chicago]. There's a certain attitude you want people to understand, walking in with a suit on."

THE SHOES
"I collect shoes like they were brushes. My suits are even dictated by the shoes. The most I spend is $200,000 on suits and shoes at one time. But I don't do that every weekend."

THE CUT
"I've always been a petite-type person, well, skinny. I hide that with oversize clothes, fabrics that drape. And I've always thought my feet were big, that great pants make your feet look smaller."

I'm **comfortable** being a **MAN**

Rockport

WWW.ROCKPORT.COM

rupaul, drag superstar.

be comfortable. uncompromise. start with your feet.

my world . . . and welcome to it?

Despite everything I've said thus far, I feel decidedly ambivalent about consumer culture's inroads into the male body. I *do* find it wonderful—as I've made abundantly clear—that the male form, both clothed and unclothed, is being made so widely available for sexual fantasy and aesthetic admiration. I like the fact that more and more heterosexual white guys are feeling permission to play with fashion, self-decoration, sensual presentation of the self. Even Dockers has become a little less "me a guy . . . duh!" in its ads and spreads for khakis, which now include spaced-out women as well as men.

But I also know what it's like to be on the other side of the gaze. I know its pleasures, and I know its agonies—intimately. Even in the second half of the twentieth century, beauty remains a prerequisite for female success. In fact, in an era characterized by some as "postfeminist," beauty seems to count more than it ever did before, and the standards for achieving it have become more stringent, more rigorous, than ever. We live in an empire ruled not by kings or even presidents, but by images. The tight buns, the perfect skin, the firm breasts, the long, muscled legs, the bulgeless, sagless bodies are everywhere. Beautiful women, everywhere, telling the rest of us how to stand, how to swing our hair, how slim we must be.

Actually, all this flawless beauty is the product of illusion, generated with body doubles, computers, artful retouching. "Steal this look!" the lifestyles magazines urge women; it's clear from the photo that great new haircut of Sharon Stone's could change a woman's life. But in this era of digital retouching not even Sharon Stone looks like Sharon Stone. (Isabella Rossellini, who used to be the Lancôme girl before she got too old, has said that her photos are so enhanced that when people meet her they tell her, "Your sister is so beautiful.") Still, we try to accomplish the impossible, and often get into trouble. Illusions set the standard for real women, and they spawn special disorders and addictions: in trying to become as fat-free and poreless as the ads, one's fleshly body is pushed to achieve the impossible.

I had a student who admitted to me in her journal that she had a makeup addiction. This young woman was unable to leave the house—not even to walk down to the corner mailbox—without a full face and body cover-up that took her over an hour and a half to apply.

In her journal, she described having escalated over a year or so from minimal "touching-up" to a virtual mask of foundation, powder, eyebrow pencil, eye shadow, eyeliner, mascara, lip liner, lipstick—a mask so thorough, so successful in its illusionary reality that her own naked face now looked grotesque to her, mottled, pasty, featureless. She dreaded having sex with her boyfriend, for fear some of the mask might come off and he would see what she looked like underneath. As soon as they were done, she would race to the bathroom to reapply; when he stayed over, she would make sure to sleep lightly, in order to wake up earlier than he. It's funny—and not really funny. My student's disorder may be one generated by a superficial, even insane culture, a disorder befitting the Oprah show rather than a PBS documentary. But a disorder nonetheless. Real. Painful. Deforming of her life.

So, too, for the eating disorders that run rampant among girls and women. In much of my writing on the female body, I've chronicled how these disorders have spread across race, class, and ethnic differences in this culture. Today, serious problems with food, weight, and body image are no longer (if they ever were) the province of pampered, narcissistic, heterosexual white girls. To imagine that they are is to view black, Asian, Latin, lesbian, and working-class women as outside the loop of the dominant culture and untouched by its messages about what is beautiful—a mistake that has left many women feeling abandoned and alone with a disorder they weren't "supposed" to have. Today, eating problems are virtually the norm among high school and college women—and even younger girls. Yes, of course there are far greater tragedies in life than gaining five pounds. But try to reassure a fifteen-year-old girl that her success in life doesn't require a slender body, and she will think you dropped from another planet. *She* knows what's demanded; she's learned it from the movies, the magazines, the soap operas.

There, the "progressive" message conveyed by giving the girls and women depicted great careers or exciting adventures is overpowered, I think, by the more potent example of their perfect bodies. The plots may say: "The world is yours." The bodies caution: "But only if you aren't fat." What counts as "fat" today? Well, Alicia Silverstone was taunted by the press when she appeared at the Academy Awards barely ten pounds heavier than her (extremely) svelte self in *Clueless*. Janeane

Garofalo was the "fat one" in *The Truth About Cats and Dogs*. Reviews of *Titanic* described Kate Winslett as plump, overripe, much too hefty for ethereal Leonardo DiCaprio. Any anger you detect here is personal too. I ironed my hair in the sixties, have dieted all my life, continue to be deeply ashamed of those parts of my body—like my peasant legs and zaftig behind—that our culture has coded as ethnic excess. I suspect it's only an accident of generational timing or a slight warp in the fabric of my cultural environment that prevented me from developing an eating disorder. I'm not a makeup junky like my student, but I am becoming somewhat addicted nowadays to alpha-hydroxies, skin drenchers, quenchers, and other "age-defying" potions.

No, I don't think the business of beauty is without its pleasures. It offers a daily ritual of transformation, renewal. Of "putting oneself together" and walking out into the world, more confident than you were, anticipating attraction, flirtation, sexual play. I love shopping for makeup with my friends. (Despite what Rush Limbaugh tells you, feminism—certainly not feminism in the nineties—is not synonymous with unshaved legs.) Women bond over shared makeup, shared beauty tips. It's fun. Too often, though, our bond is over shared pain—over "bad" skin, "bad" hair, "bad" legs. There's always that constant judgment and evaluation—not only by actual, living men but by an ever-present, watchful cultural gaze which always has its eye on our thighs—no matter how much else we accomplish. We judge each other that way too, sometimes much more nastily than men. Some of the bitchiest comments about Marcia Clark's hair and Hillary Clinton's calves have come from women. But if we are sometimes our "own worst enemies," it's usually because we see in each other not so much competition as a reflection of our fears and anxieties about ourselves. In this culture, all women suffer over their bodies. A demon is loose in our consciousness and can't easily be controlled. We see the devil, fat calves, living on Hillary's body. We point our fingers, like the accusers at Salem. Root him out, kill *her*!

And now men are suddenly finding that devil living in their flesh. If someone had told me in 1977 that in 1997 *men* would comprise over a quarter of cosmetic-surgery patients, I would have been astounded. I never dreamed that "equality" would move in the direction of men worrying *more* about their looks rather than women worrying less. I first suspected that something major was going on when the guys in my gender classes stopped yawning and passing snide notes when we discussed body issues, and instead began to protest when the women talked as though they were the only ones "oppressed" by standards of beauty. After my book *Unbearable Weight* appeared, I received several letters from male anorexics, reminding me that the incidence of such disorders among men was on the rise. Today, as many as a million men—and eight million women—have an eating disorder.

Then I began noticing all the new men's "health" magazines on the newsstands, dispensing diet and exercise advice ("A Better Body in Half the Time," "50 Snacks That Won't Make You Fat") in the same cheerleaderish mode that Betty Friedan had once chastised the women's magazines for: "It's Chinese New Year, so make a resolution to custom-order your next takeout. Ask that they substitute wonton soup for oil. Try the soba noodles instead of plain noodles. They're richer in nutrients and contain much less fat." I guess the world doesn't belong to the meat-eaters anymore, Mr. Ben Quick.

It used to be a truism among those of us familiar with the research on body-image problems that most men (that is, most straight men, on

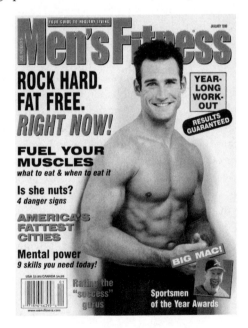

whom the studies were based) were largely immune. Women, research showed, were chronically dissatisfied with themselves. But men tended, if anything, to see themselves as better-looking than they (perhaps) actually were. Peter Richmond, in a 1987 piece in *Glamour*, describes his "wonderful male trick" for seeing what he wants to see when he looks in the mirror:

> I edit out the flaws. Recently, under the influence of too many Heinekens in a strange hotel room, I stood in front of a wraparound full-length mirror and saw, in a moment of nauseous clarity, how unshapely my stomach and butt have become. The next morning, looking again in the same mirror, ready to begin another business day, I simply didn't see these offending areas.

Notice all the codes for male "action" that Richmond has decorated his self-revelation with. "Too many Heinekens," "another business day"—all reassurances that other things matter more to him than his appearance. But a decade later, it's no longer so easy for men to perform these little tricks. Getting ready for the business day is apt to exacerbate rather than divert male anxieties about the body, as men compete with fitter, younger men and fitter, more self-sufficient women. In a 1994 survey, 6,000 men ages eighteen to fifty-five were asked how they would like to see themselves. Three of men's top six answers were about looks: attractive to women, sexy, good-looking. Male "action" qualities—assertiveness, decisiveness—trailed at numbers eight and nine.

"Back when bad bodies were the norm," claims *Fortune* writer Alan Farnham (again, operating with the presumption of heterosexuality), "money distinguished male from male. Now muscles have devalued money," and the market for products and procedures "catering to male vanity" (as *Fortune* puts it) is $9.5 billion or so a year. "It's a Face-Lifted, Tummy-Tucked Jungle Out There," reports *The New York Times*. To compete, a man

> could buy Rogaine to thicken his hair. He could invest in BodySlimmers underwear for men, by the designer Nancy Ganz, with built-in support to suck in the waist. Or he could skip the aloe skin cream and go on to a more drastic measure, new to the male market: alpha-hydroxy products that slough off dead skin. Or he could rub on some belly- and thigh-shrinking creams . . . If rubbing cream seems too strenuous, [he] can just don an un-

dershirt from Mountainville House, to "shape up and pull in loose stomachs and sagging chests," with a diamond-shaped insert at the gut for "extra control." . . . Plastic surgery offers pectoral implants to make the chest appear more muscular, and calf muscle implants to give the leg a bodybuilder shape. There is liposuction to counter thickening middles and accumulating breast and fatty tissue in the chest . . . and a half-dozen surgical methods for tightening skin.

Some writers blame all this on sexual equality in the workplace. Anthropologist Lionel Tiger offers this explanation: "Once," he says, "men could fairly well control their destiny through providing resources to women, but now that the female is obliged to earn a living, he himself becomes a resource. He becomes his own product: Is he good-looking? Does he smell good? Before, when he had to provide for the female, he could have a potbelly. Now he has to appear attractive in the way the female had to be." Some evidence does support this. A *Psychology Today* survey found that the more financially secure the woman, the more important a man's looks were to her.

I, however, tend to see consumer capitalism rather than women's expectations or proclivities as the true motor driving male concern with appearance. Calvin gave us those muscled men in underwear. Then the cosmetics, diet, exercise, and surgery industries elbowed in, providing the means for everyone to develop that great Soloflex body. After all, why should they restrict themelves to female markets if they can convince men that their looks need constant improvement too? The management and enhancement of the body is a gold mine for consumerism, and one whose treasures are inexhaustible, as women know. Dieting and staving off aging are never-ending processes. Ideals of beauty can be endlessly tinkered with by fashion designers and cosmetic manufacturers, remaining continually elusive, requiring constant new purchases, new kinds of work on the body.

John Berger's opposition of "acting" and "appearing," this body work reveals, is something of a false duality—and always has been. "Feminine" attention to appearance is hardly the absence of activity, as men are learning. It takes time, energy, creativity, dedication. It can *hurt*. Nowadays, the "act/appear" duality is even less meaningful, as the cultivation of the suitably fit appearance has become not just a matter of sexual allure but also a demonstration that one has the "right stuff": will, discipline, the ability to stop whining and "just do it." When I was growing up in the sixties, a muscular male body meant beefy but dumb jock; a middle-class girl could drool over him but probably wouldn't want to marry him. Today, with a booming "gymnasium culture" existing (as in ancient Greece) for professional men and with it a revival of the Greek idea that a good mind and a good body are not mutually exclusive, even Jeff Goldblum has got muscles, and the only type of jock he plays is a computer jock.

All of this, as physicians have begun to note, is landing more and more men straight into the formerly female territory of body-image dysfunction, eating disorders, and exercise compulsions. Last year, I read a survey that reported that 90 percent of male undergraduates believe that they are not muscular enough. That sent warning bells clanging in my mind, and sure enough, there's now a medical category for "muscle dysmorphia" (or "bigorexia," as it's actually sometimes called!), a kind of reverse anorexia in which the sufferer sees his muscles as never massive enough. Researchers are "explaining" bigorexia in the same dumb way they've tended to approach women's disorders—as a combination of bad biochemistry and "triggering events," such as being picked on. They just don't seem to fully appreciate the fact that bigorexia—like anorexia—only blooms in a very particular cultural soil. Not even the ancient Greeks—who revered athletic bodies and scorned weaklings, but also advised moderation in all things—produced "muscle dysmorphics." (Or at least, none of the available medical texts mention anything like it.) Anorexia and bigorexia, like so many contemporary disorders, are diseases of a culture that doesn't know when to stop.

Those beautiful bodies of Greek statues may be the historical inspiration for the muscled men in underwear of the Calvin Klein ads. But the fact is that studying the ancient Greeks reveals a different set of attitudes toward beauty and the body than our contemporary ideals, both homosexual and heterosexual. As is well known by now (although undiscussed when I studied philosophy as an undergraduate), Plato was not above appreciating a beautiful young body. In *Symposium*, he describes the beauty of the body as evidence of the presence of the divine on earth, and the original spur to all "higher" human endeavors (as well as earthly, sexual love). We see someone dazzling, and he or she awakens the soul to its natural hunger to be lifted above the mundane, transitory, mortal world. Some people seek that transcen-

dence through ordinary human intercourse, and achieve the only immortality they will know through the begetting of human offspring and
the continuation of the human race. For others, the beautiful body of
another becomes the inspiration for a lifelong search for beauty in all its
forms, the creation of beautiful art, beautiful words, beautiful ideals,
beautiful cities. They will achieve their immortality through commu-
nion with something beyond the body—the idea of Beauty itself.

So human beauty is a pretty far-ranging and powerful thing for
Plato, capable of evoking worlds beyond itself, even recalling a previous life when we dwelt among timeless, perfect forms. But human
beauty, significantly (in fact, all earthly beauty), can only offer a
glimpse of heavenly perfection. It's our nature to be imperfect, after all,
and anyone who tries to overcome that limitation on earth is guilty of
hubris—according to the Greeks. Our own culture, in contrast, is one
without "limits" (a frequent theme of advertisements and commercials) and seemingly without any fear of hubris. Not only do we expect
perfection in the bodies of others (just take a gander at some personal
ads), we are constantly encouraged to achieve it ourselves, with the
help of science and technology and the products and services they
make available to us. "This body could be yours," the chiseled Greek
statue in the Soloflex commercial tells us (and for only twenty minutes
three times a week—give me a break!). "Timeless Beauty Is Within
Your Reach," reads an ad for cosmetic surgery. Plato is rolling over in
his grave.

For Plato (unlike Descartes) there are no "mere" physical bodies;
bodies are lit with meaning, with memory. Our culture is more Cartesian; we like to think of our bodies as so much stuff, which can be
tinkered with without any consequences for our soul. We bob our
"family noses," lift our aging faces, suction extra fat, remove minor
"flaws" with seemingly little concern for any "deep" meaning that our
bodies might have, as repositories of our histories, our ethnic and
racial and family lineage, our personalities. Actually, much of the time
our intentions are to deliberately shed those meanings: to get rid of
that Jewish nose, to erase the years from our faces. Unlike the Platonic
philosopher, we aren't content to experience timelessness in philosophy, art, or even the beautiful bodies of others; we want to stop time
on our own bodies too. In the process, we substitute individualized
beauty—the distinctive faces of the generation of beautiful actresses of

my own age, for example—for generic, very often racialized, reproducible codes of youth.

The fact is that we're not only Cartesian but Puritan in our attitudes toward the body. The Greeks went for muscles, sure, but they would have regarded our exercise compulsions as evidence of a system out of control. They thought it unseemly—and a failure of will—to get too self-obsessed with *anything*. They were into the judicious "management" of the body (as French philosopher Michel Foucault has put it), not its utter subjugation. We, on the other hand, can become what our culture considers to be sexually alluring only if we're willing to regard our flesh as recalcitrant metal, to be pummeled, burned, and tempered into steel, day in and day out. No pain, no gain. Obsessively pursuing these ideals has deprived both men *and* women of the playful eros of beauty, turned it all into constant, hard work. I love gay and black body cultures for their flirtatiousness, their tongue-in-cheekness, their irony, their "let's dress up and have some fun" attitudes. Consumer culture, unfortunately, can even grind playfulness into a commodity, a required item for this year's wardrobe.

For all its idealization of the beauty of the body, Greek culture also understood that beauty could be "inner." In the *Symposium*, a group of elite Greeks discourse on the nature of love. Everyone except for Socrates and Aristophanes is in love with someone else at the party, and they're madly flirting, advancing their own romantic agendas through their speeches. Among the participants are the most beautiful young men of their crowd. Socrates himself is over fifty at the time, and not a pretty man to look at (to put it generously). Yet as we're told at the beginning (and this seems to have been historically true), nearly everyone has at one time or another been "obsessed" with him, "transported, completely possessed"—by his cleverness, his irony, his ability to weave a spell with words and ideas. Even the most dazzling Athenian of them all—soldier superhero Alicibiades, generally regarded as one of the sexiest, handsomest men in town, who joins the party late (and drunk) with a beautiful wreath of violets and ivy and ribbons in his hair—is totally, madly smitten with Socrates.

Alcibiades' love for Socrates is *not* "Platonic" in the sense in which we have come to understand that term. In fact, Alcibiades is insulted because Socrates has refused to have sex with him. "The moment he starts to speak," he tells the crowd of his feelings for Socrates, "I am beside myself: my heart starts leaping in my chest, the tears come streaming down my face." This is not the way it usually goes. In the more normal Greek scheme of things, it's the beautiful young man—like Alcibiades—who is supposed to start the heart of the older man thumping, and who flirtatiously withholds his favors while the older lover does his best to win him. Alcibiades is in a state about this role reversal, but he understands why it has happened. He compares Socrates to a popular kind of satyr statue, which (like the little lacquered Russian dolls we're more familiar with) could be opened to reveal another figure within. Socrates may be ugly as a satyr on the outside, but "once I had a glimpse of the figures within—they were so godlike, so bright and beautiful, so utterly amazing, that I no longer had a choice—I just had to do whatever he told me."

We pay constant lip service to beauty that is more than skin-deep. The talk shows frequently parade extreme May-December matings for our ogling too. But the fact is that the idea of a glamorous young man being romantically, *sexually* obsessed with someone old and "ugly"—same-sex or other-sex and no matter what other sterling qualities he or she may have—is pretty much beyond us. Historically, men have benefited from a double standard which culturally codes their gray hair, middle-age paunches, facial lines, as signs of wisdom and experience rather than advancing decrepitude. My older gay male friends lament that those days are over for them. And if those new polls about women's attitudes are to be believed, the clock is ticking on that double standard for heterosexual men, too—no matter how hard Hollywood tries to preserve it. With more and more expectation that men be as physically well-tended as women, those celluloid pairings of Woody Allen and women half as old and forty-six times as good-looking are becoming more of a hoot every day.

There is something anti-sensual to me about current aesthetics. There's so much that my younger friends go "uggh" over. Fat—yecch! Wrinkles—yuck! They live in a constant state of squeamishness about the flesh. I find that finely muscled young Calvin Klein model beautiful and sexy, sure. But I also was moved by Clint Eastwood's aging chest in *The Bridges of Madison County*. Deflated, skin loose around the waistband of his pants, not a washboard ridge in sight—for me, they signaled that Eastwood (at least for this role) had put Dirty Harry away for good, become a real, warm, penetrable, vulnerable human being instead of a make-my-day machine. Call me old-fashioned, but I find that very sexy. For a culture obsessed with youth and fitness, in contrast, sagging flesh is almost the ultimate signifier of decay and disorder. We prefer the clean machine—and are given it, in spades. Purified of "flaws," all loose skin tightened, armored with implants, digitally enhanced, the bodies of most movie stars and models are fully dressed even when naked.

In *Saturday Night Fever*, John Travolta had been trim, but (by contemporary standards) a bit "soft." Six years later, Travolta re-created Tony Manero in the sequel, *Staying Alive*. This time, however, the film was directed by Sylvester Stallone, who showed Travolta a statue of a discus thrower and asked, "How would you like to look like that?" "Terrific," Travolta replied, and embarked on a seven-month program of fitness training that literally redesigned his body into a carbon copy of Sly's. In the film, his body was "perfect": gleaming and muscular, without an ounce of fat. He was nice to look at. But if I had to choose between the Tony Manero of *Fever* and the Tony Manero of *Staying Alive*, it'd be no contest. I'd rather spend time (and have sex) with a dancing man with love handles than with a Greek statue who gets in a nasty mood if he misses a workout.

the male animal, reconsidered

gentleman or beast?

THE DOUBLE BIND OF MASCULINITY

the "hot man" thesis

Officially, our culture is supposed to be miles beyond the Victorian notion that men are bundles of raging animal instincts, while women are the sweet, pure, sexless guardians of civilization who keep the brutes at bay. My generation of feminists are continually chastised for holding that view (although few actually do), which Naomi Wolf dubbed "victim feminism," joining Katie Roiphe, Camille Paglia, and others in urging that feminists stop blaming men for date rape and sexual harassment and start acknowledging women's active, initiating sexual role. But the truth is that feminists are the scapegoats here. All you have to do is take a quick look around, and it's clear that the days of our (very short-lived) cultural love affair with the idea that men and women are basically of the same species, with similar drives and desires, are over, and it's not feminists who turned the romance sour.

While feminists have been writing tomes on how diverse, fluid, and fragmented sexual identity is, "the new science of the brain" (as *Newsweek* calls it) is declaring differences between men and women to be hard-wired. While we've been sitting at our computers, hotly churning out analyses of female desire, popular science—both "hard" and "social"—has been busy reestablishing that men are testosterone-driven, promiscuous brutes whom nature won't permit to keep their peckers in their pants. I knew we had gone very, very far in the direction of a new biologophilia when scientist Steven Pinker, commenting on Bill Clinton's libido, presented "an evolutionary explanation for Presidents behaving badly." The explanation begins with a version of what has become standard-issue popular science nowadays. Simply put, the idea is that while females have a tremendous amount invested, evolutionarily speaking, in the survival and success of each individual egg, males get greater returns on multiple inseminations. Pinker writes:

"A prehistoric man who slept with fifty women could have sired fifty children, and would have been more likely to have descendants who inherited his tastes. A woman who slept with fifty men would have had no more descendants than a woman who slept with one. Thus, men should seek quantity in sexual partners; women, quality—a source of protection, resources, and good genes for their children."

Indeed, Pinker goes on to point out, "in all societies known to ethnography it is the males who seduce, proposition, hire prostitutes, and accumulate spouses." What makes presidential (and celebrity) male libido so much more efficient at all this is simply greater opportunity: "[F]or the average Joe the issue is moot: he won't find eight women willing to sleep with him in the next two years, and he never has to worry about attractive women propositioning him out of the blue," while if a man is sufficiently rich, handsome, or powerful, he can collect a harem—and is faced with the temptation to do so all the time. Perhaps, too, Pinker speculates, some especially high-powered male brains are hard-wired for greater risk-taking than the average male.

It's not only evolutionary science that has been advancing these notions. Versions of the "hot man" thesis have been cropping up all over our culture, from pop psychology to television. The pump was already primed by John Gray, best-selling author, sex therapist, and infomercial-meister, who's made millions by advising couples to look at each other as descendants of radically different galactic species. "The next time you are frustrated with the opposite sex," writes Gray, "remember that men are from Mars and women are from Venus." It's a metaphor, of course, but one that Gray extends quite rigorously and colorfully, drawing on many secondary metaphors to describe the incommensurable habits and personalities of our two alien species. For example: Men are like rubber bands ("he needs to pull away before he can get closer"), while women are like waves (our self-esteem naturally rises, falls, and then rises again, like a tide obeying the moon). Gray's metaphor of different planetary origins has so infiltrated popular culture that writers just lazily grab at it whenever they want to say that males and females behave differently. "What some researchers are finding," write the authors of a 1998 *Newsweek* piece (called "Boys Will Be Boys"), "is that boys and girls really are from two different planets. But since the two sexes have to live together here on Earth, they should be raised with special consideration for their distinct needs." Hmmm, kind of a weird cosmogony. I guess we were strapped into our pink and blue spaceships at birth, and . . . rockets away!

Unlike the brain scientists and biologists, Gray doesn't claim to know the origins of our differences, and admits that he doesn't really care whether the behavior he describes is socially constructed or biologically rooted. He is, rather, a phenomenologist of sexual differences as he finds them and—although multiculturalism seems to have utterly passed him by (there's not a word about racial or ethnic differences in his book)—he is often quite astute. My sophisticated poststructuralist graduate students, who normally believe that to generalize about gender is the worst sin one can commit, adored his stuff on men needing to go to their caves periodically; every woman in the room knew what he was talking about. We all got a great laugh too at his recommendation that relationships could be saved if only men would say three little words when they feel the need to retreat: "I'll be back." (Gray, apparently not having gone to a movie in years, makes this recommendation without irony. Nonetheless, it's a good one!)

But like contemporary pop science, Gray also seems a bit drunk on sexual difference. I don't think there's a line in any of his books that suggests that men and women share a single proclivity, reaction, fantasy, or need. "Even when women think they are from Mars [career women, for example, with "male" commitment problems], they are still from Venus," Gray insists, and are in deep danger unless they begin catering to their basic Venusian nature. Reading what he has to say about that nature is sometimes like traveling backward in time to about 1955, particularly when Gray is talking about sex. In *Mars and Venus on a Date*, he likens male sexuality to a blowtorch ("heats up really fast and then turns off in an instant"), while women are like (conventional, not microwave) ovens—we "slowly heat up and slowly cool down." Example: "A man may go out of town for a week. When he gets back, the first thing he wants to do is have sex, but his wife may not feel the same way. Her feeling is 'How can you just want to have sex? We haven't talked for days. Don't you even care about how I feel?' " You see, we need to "warm up" gradually. And I suppose that during periods of separation from our lovers, too, women-ovens (apparently lacking even a pilot light) simply go stone-cold—engage in no fantasies, experience no longings, feel no sexual frustration that might keep the pot even slightly simmering.

Women need to *talk* in order to raise their sexual temperature, Gray argues, because unlike men, we get warmed up not by physical but by mental chemistry. When we first meet someone we're attracted to, the attraction is due to something about the man's character (or something we imagine about his character). In contrast, man's blowtorch is ignited by "mindless" (Gray's word, not mine) physical attraction to body parts: "her hair, her smile, her eyes, her height, her legs, her rear, her breasts," or "the overall shape of her body." "His attraction has nothing to do with who this woman is . . . He only knows he wants to see more, touch more, and feel more." (Hey, I've felt that way! And acted rather recklessly on those feelings, more times than I care to remember. I guess I've forgotten my planetary origins. Maybe my testosterone levels are too high. I haven't invested enough in my eggs! God, what's wrong with me?*)

Putting the evolutionary argument and Gray's metaphors together, the idea of the female "oven" does double duty, as the place where those precious genetic buns are baked—hopefully, only after the equipment has been warmed up. With those blowtorches, though, you can't count on a slow ignition; if the fire is lit, they're just as likely to rape you as woo you. Evolutionary biologist Randy Thornhill of the University of New Mexico believes this, arguing that aggressive male sexuality, a sexuality which couldn't give a damn about what the female it mates with wants, has been favored by natural selection. "Essentially,"

---

*Unlike Gray, I'm inclined to think that blowtorch-style passion is ignited by finding oneself in the position of the pursuer, rather than being written in our gender makeup. Desire feeds on the perception of need, want, lack; so long as women are encouraged (as Gray suggests) to always be the withholding object of male need, of course men are going to be hot for us and we're going to feel somewhat cooler toward them. My sexuality has never felt so "male" (as Gray would label it) as on those occasions when I was in the presence of someone I was interested in but (for one reason or another) unable to "have."

he says, "all males, in the right circumstances, respond to forced sex."
(You see what I mean about feminists?)

Descending still further down the ladder of scientific legitimacy, similar ideas have lately begun to crop up in a raft of purportedly humorous little advice books, usually written by yuppie white men,
generously sharing the secrets of male psychology with women. *What
Men Want: Three Professional Single Men Reveal What It Takes to
Make a Man Yours* advises women to keep their men on a very short
leash, because "a man will be polygamous unless something or someone restrains his natural inclination," and unless inhibited by fear
"will run after pleasure wherever he can find it." "That's the ugly
truth," the authors resignedly (or proudly) admit. They claim to be
helping women out by giving them the Real Poop on the male animal,
so we will be properly armed with the information needed to manipu-
late the men in our lives. They accept as a given the "scarcity thesis":
that marriageable men—particularly lawyers and doctors such as
themselves (there is a photo of the three of them dressed for success on
the back cover)—are a resource so precious that hours of housebreak-
ing (and self-grooming) will seem a slim price to pay for snagging one
of your own.

Though they purport to help women, an ugly, aggressive misogyny
runs throughout these books, as it does in *Esquire*'s frequent articles
("The Post-Sensitive Male," "The Return of the Alpha Male," etc.)
challenging men to "abandon the feminized *faux male*" ideal—"Mr.
Sensitive"—foisted on them by women, and demand the "freedom to
be guys again," to no longer allow women to "dictate the terms of the
battle of the sexes." The pieces rouse men to assert themselves, to acknowledge that they're tired of pretending, tired of being asked to
change, to become more "thoughtful," more "relationship-oriented."
They encourage men to stick their heads (either one) out the window
and shout to women that they're mad as hell and aren't going to take
it anymore. Men are sex-crazed jerks, and you're just going to have to
live with it!

Another in this genre is *What Men Don't Want Women to Know:
The Secrets, the Lies, the Unspoken Truth*. The central premise of this
book is that "man spends his life in one of two basic states: loaded or
unloaded." When a man's penis is loaded (the authors advise us to

"think of the penis as a gun"), he will do everything he can to unload it, so a woman is well-advised to give her man an orgasm before she sends him off to work in the morning. We are also told that "if men could get away with it, they would buy and sell women like slaves," that "men lie constantly," and "in the eyes of your man, there can never be too many women who lust for and want him." If these guys are right (and don't forget, the biologists are saying much the same thing in more scientific, less misogynist language), an American President was brought to the edge of expulsion from office for . . . being a man!

"boys will be boys": the cost to girls and boys

Could it be that our culture has a small problem knowing what it wants from men? Think of the instruction in raw aggression that football provides and how it encourages the player to think of his body as a fierce, unstoppable force of nature. Think of how this aggression is rewarded—with scholarships, community adulation, romantic attention, special attendance deals cut with teachers, administrative leniency when "boys will be boys." Now imagine the young quarterback at a workshop on date rape, held by the counseling center of the same high school which is encouraging him to be an animal on the football field. At that workshop, he's told he must learn that he is not an animal, that his body is not an unstoppable force of nature, that it must yield, in fact, to one little word. Now, which is this young man supposed to be . . . an animal or a gentleman?

The answer, of course, is that he is supposed to be both. But that's not so easy—except in the world of cultural fantasy. Here, I return to the subthemes of that *Ally McBeal* "big penis" episode which I discussed earlier in this book. In one of those subthemes, the firm defends a young man who broke a guy's cheekbone when he called his girl a slut. As the defense begins, cute li'l Ally is up first, and argues (with not much conviction) that the young man's punch was the "chivalrous" thing to do. "I'm single. I date," she tells the jury (the memoir form is big among the lawyers on this show) and *she* would want her date to take action if her "chastity was impugned." The whiff of the nineteenth century is in the air. It's not yet overpowering, however, since we know that Ally, although she wears adorable oversize pajamas, is hardly a virgin; in fact, she's "in heat"—as she puts it—in this particular episode.

But then the smartest, albeit most neurotic, male member of the firm (played by Peter MacNicol)—who's just about as little and cute as Ally, although it works less to his romantic advantage—takes his turn at bat and argues a different line—that the punch was "man's human nature." "Every man is part warrior," he tells the jury; those "primal qualities will always be there." He recounts an episode in which some guy bumped into him rudely and he didn't stick up for himself. His lack of action haunted him for weeks afterward. The next time it happens, he knocks the guy cold. (It's hard to imagine, but never mind.) "Am I embarrassed? Yes," he tells the jury. "But as a man, the most satisfying moment in my life was that punch." Turning now to the young man he's defending, he urges the jury to look deep into their souls and "Admit it . . . You're glad he threw that punch."

The script not only has the jury "admitting it," but heroine Ally too. This is where the show goes beyond those backlash books and articles which proclaim "The Return of the Alpha Male," while telling women to like it or lump it. Ally's message for women is *not*: "Men are animals; if you want one of them, you're just going to have to learn to live with it," but "Men are animals. And ain't it just grand!" Strolling with her large-membered date the night of what she refers to as the firm's legal "victory for Cro-Magnon man," she confesses that if anyone insulted her, she would want her date to "rip his head off." If he just turned the other cheek and walked away, she'd be "disappointed." It has little to do with chivalry, however (as the "Cro-Magnon" metaphor underscores). It has to do with the magnetic appeal of "uncivilized," untamed maleness.

The same episode "vindicates" the appeal of boxing—importantly, to women as well as men, showing a mixed-sex group smoking cigars and wildly cheering as two men beat the shit out of each other in the ring. At the same time, it crosscuts to scenes of Ally having earth-moving sex with the well-hung model. As one fighter is knocked down and declared "out," we cut to a scene of Ally sleeping peacefully, post-coitum, equally knocked out. Primal punches and primal sex. It's not even that a girl has to tolerate the first in order to get the second. The argument of the "big penis/rip his head off" episode (ouch) is not really that "size matters," it's that primal nature matters. The primitive male animal—if we're honest with ourselves—turns a girl on. In the show, admitting to the joys of the big penis and being rewarded with those very joys is really just a metaphor for that acknowledgment.

While equating them symbolically, the show conveniently keeps the knockout punches and the knockout sex in two different rooms. That's the lie the show tells, that a culture can celebrate untamed male aggression and keep it nicely, sexily contained at the same time. The two men go at each other like wild beasts in the ring, while Ally has passionate sex with the model. But her model is anything *but* a wild beast; he's so respectful of her space that he leaves before morning, because *she* wants him to. He may be capable of bringing out the primal urges in Ally, sure, but not because he's a thug who punched some other guy out. He's adorable, he's kinda laid back, and he's got a giant member. That would bring out my primal urges too. But what does any of that have to do with the magnetic appeal of the "warrior male"?

I hate to tell you this, Ms. McBeal, but the *real* "warrior males"— those guys ripping each other's heads off in the ring—might be a bit different on a date. I'm not saying that there aren't boxers—like Evander Holyfield, for example—who seem to be gentle, sweet men outside the ring. Clearly, there are. I'm saying that we really oughtn't to be surprised that boxers, basketball players, and football players (from high school on up to the pros) get charged with rape and public vandalism far more often than any other group of financially and socially successful males. When Mike Tyson raped Desiree Washington in 1992, Joyce Carol Oates wrote a brilliant piece for *Newsweek* in which she noted that boxing "excessively rewards men for inflicting injury upon one another that, outside the ring, with less 'art,' would be punishable as aggravated assault, or manslaughter." As spectators, we find these "displays of masculine aggression" exciting in the ring precisely *because* they break with the taboos of civilization, act out the (forbidden) aggression in all of us. Indeed, we *want* the boxer to be as uncivilized as he can be; we reward him for it. In the meantime the boxer, "conduit" for all this suppressed aggression, vestigial repository of primal masculinity, is a real person who is learning—in the very fibers of his being, his body—that civilized taboos against violence do not apply to him.

At the same time as violence is being bred in the bone of our primitive gladiators, the tremendous cultural adoration heaped on them—by men and women alike—encourages a sense of absolute entitlement that doesn't combine very well with respect for property rights or women's wishes. I made the next-to-biggest mistake of my life as a college teacher at a small liberal arts college when I allowed one of my students, who was friends with some of the biggest (physically and in terms of success) athletes on the neighboring university football team, to invite his pals to our end-of-class party—held on college property. Then I made my biggest mistake: I left the party early, leaving the women and the room at the mercy of these guys. I had left my students to continue a class party without me before, with no problems. This time it was different. No one was raped, but several women were badly harassed, and the room was completely trashed, furniture broken and beer soaked through the carpet. When the incident became known, the biggest reaction of friends and colleagues was shock at my naiveté. That's the way football players party, I was told.

The cultural adoration of those athletes who "perform" masculinity for us often continues even after they have been charged with or convicted of serious crimes. We now know that the Los Angeles police, far from conspiring against O. J. Simpson, treated him so deferentially that they seriously botched their initial interview with him. (If there was a conspiracy at work, it was probably the unspoken gender solidarity that had allowed the police, during the years before the murders, to ignore or minimize Nicole's many calls for help.) While Simpson was in jail during the trial, he and police deputies frequently palled around "like guys hanging out on a street corner," as Shawn Chapman, a black female attorney who was working behind the scenes on the defense team, put it. She described to Lawrence Schiller a conversation she overheard as a "pretty alternate juror" walked by, dressed in Levi's:

"Check out Fufu in jeans," one deputy cracks.

"Oh, man, look at her in those pants." Simpson is leering. "I want to get *at* her."

"Yeah, she's hot," another deputy agrees, "but she's so stupid."

"I don't care how stupid she is," Simpson says. "Look at her in those jeans!"

After Mike Tyson was convicted of rape, a group of black ministers in Indiana circulated petitions calling for Tyson to be given a suspended sentence, arguing that he was a hero to African-American youth (I guess they meant African-American *boys*), and that it would cause them great pain to "see a fallen hero"; shortly before the verdict came down, Donald Trump offered to pay off Desiree Washington and the state of Indiana if Tyson were shown "mercy." My belief is that, to this day, Tyson himself doesn't believe he did anything seriously wrong that night. Nor did the teenage athletes from the prosperous New Jersey suburb of Glen Ridge consider that they did anything wrong in 1989 when they lured a developmentally disabled girl into the basement of one of their homes, made her perform oral sex on one of them, and then proceeded to rape her with a broom handle and baseball bat.

The Glen Ridge community was football-crazy, and idolized the members of the high-school team. "I had never seen anything like it before," Susan Atkins, who was a teacher at the high school for twenty-four years, told Bernard Lefkowitz, whose extraordinary *Our Guys* chronicles and brilliantly analyzes the Glen Ridge rape. "Football was king and if you didn't wear a football uniform you were a nobody." If you did wear that uniform, however, you could count on being able to get away with anything. Several of the rapists, in the years prior to the rape, had been allowed to get away with harassment, exhibitionism, and severe vandalism, all in the name of "boys will be boys"—or more precisely, jocks will be jocks. When one of them led five of his buddies in a guerrilla raid on the country club, tearing up the tennis courts, driving golf carts into the walls of the garage, vandalizing the locker room, the club was quietly paid off and pressed no charges. With girls, one of their favorite group activities was "voyeuring"—one guy would get a girl to perform oral sex while a group of other guys watched from a closet or a hall, their presence unknown to the girl. The practice was seen, even by other girls, as the fault of the girl herself, for having been so gullible or stupid. "The way Tara saw it," Lefkowitz reports about the attitude of a popular cheerleader, "you couldn't hold the guys responsible for what happened; you had to expect a high school guy to act crudely. It was up to the girl to control the situation."

When the boys were arrested, the entire community—including coaches, teachers, the principal, and the school board, as well as parents—rallied around them just as the ministers had rallied around Tyson. These affluent white businesspeople were more successful, however, in winning "mercy" for their boys. Four of the young men were convicted of rape by a jury that reflected the racial, urban, and suburban blend of Essex county (eight whites, eight blacks, and a mix of economic and professional backgrounds). They could have been sentenced to as much as forty years. The judge, however, effectively wiped out that guilty verdict, sentencing the young men (who were seventeen when they committed the crime) to a "young adult offender's institution" with no minimum term of imprisonment, and then—virtually unprecedented for defendants convicted of first-degree sex crimes, and certainly unlike the treatment of youthful, urban, black offenders—allowing them to go free on bail pending the outcome of their appeals. Today, nearly ten years later, those appeals are still not decided, and not one of the young men has yet seen the inside of a jail cell.

The notion that "boys will be boys" may work to male advantage when it comes to privileged high school and college athletes or famous celebrities who commit sex crimes. But believing that it is in male "nature" to be aggressive and sexually trigger-happy is a double-edged sword. As clinical psychologist William Pollack points out, it creates a picture of males as potentially "toxic" that hardly benefits most ordinary boys:

"When a seven-year-old boy impulsively plants a kiss on a somewhat unwilling female playmate, he is branded a sexual delinquent and suspended from school. A fifth-grade boy coming directly from a 'sex education' class jokes with a girl that her sagging belt looks like a penis and gets accused of 'sexual harassment.' . . . Put aside for the moment the obvious double standard of how teenage girls who joke about the 'bulge' in football players' pants are unlikely to be branded as sexual harassers, and ask why we have confused boys' childish exploratory play with adult predatory behavior. No doubt some components of boys' and girls' play go beyond the bounds of acceptability and are deserving of redirection or even reprimand. Yet, when it is a *boy* involved, we seem to forget his need to play, experiment, and fail

in order to grow. Instead, we respond as though he is a full-fledged aggressor."

What a quandary all this places boys in! On the one hand, the official and unofficial wisdom is declaring that to be a "real male" (you
know, the kind that comes from Mars not Venus) *requires* that one be
fiercely competitive and aggressive in "manly arenas" like athletics
and have a predatory, promiscuous, get-all-you-can attitude toward
girls, and woe to the boy who doesn't. He'll likely be subject to taunting, goading, accusations of wimpiness, and probably gay-bashing—
whether he's gay or not. "Anything tender, anything compassionate, or
too artistic is labeled as gay," says psychologist Michael Thompson,
who writes about boys, "The homophobia of boys in the 11, 12, 13
range is a stronger force than gravity." On the other hand, our little
Martian knows that if he makes the wrong move with a girl, he may
be in deep trouble.

In Glen Ridge, the young athletes didn't worry about getting into
trouble, because they never did. They had nothing but contempt for
girls,* so they certainly didn't worry about offending *them*. It was their
fathers and coaches whom the boys cared about:

"Their clothes and uniforms, their jibes and jokes . . . were elements
in a rehearsal for manhood—or that special definition of manhood
they picked up from their fathers and brothers. By that definition, if
you weren't a winner, if you walked around with the hangdog expression of a loser, you were less than a man. Kids who weren't versed in
jock talk, kids who didn't run out a ground ball or who swung their
bats lethargically, kids who carried a book with them to Carteret
would be labeled 'queers' or 'nerds.' These kids would never be real
men."

Clearly, all boys pay in one way or another by being encouraged to
conform to such standards of manhood. The cruelest injustice, however, is that at the same time as the "winners" may be getting away
with sexual harassment and rape, the "nerds" and "queers" are paying

---

*Lefkowitz describes their vocabulary: "A girl who gave them oral sex was 'hoovering'—like a
vacuum cleaner . . . They would often refer to girls who were available as 'animals.' A girl who
gave blow jobs to a group of guys one after another was nicknamed 'Seal.' This meant . . . that
the girl was behaving 'like a trained seal in the circus, doing whatever they commanded.' When
they saw one of the marks coming toward them from across the room, the guys would put on
their beer goggles and chant: 'Oink, oink, oink.' "

the price for the bad reputation *all* boys have been tarred with. The thoughtful, intelligent, caring boy—one who is least likely to behave disrespectfully with a girl—may find himself paralyzed, caught between his desire to act "like a man" and fear of giving offense. Pollack quotes Mitchell, a teenager telling his father about his first date and how uncomfortable it was:

"[W]hen I was driving her home we kind of ran out of conversation. I think she was expecting me to attack her or something . . . The whole good-night kiss thing, you know. But, sometimes, I feel like touching a girl is illegal, if she doesn't want you to. I wasn't sure if she wanted me to. So, then I started thinking that if I tried to kiss her she'd hit me with a sexual harassment suit . . . I was so confused I couldn't even figure out if I really wanted to or not. So, I figured I'd better just say good night and drop her off. Now she probably thinks I'm a wimp, or that I don't like her."

The ironic—and very sad—thing here is that Mitchell's date was probably only trying to conform to *her* proper dating etiquette as a Venusian. For, as Gray informs us, "Men Pursue and Women Flirt." "To be most attractive," he writes, "a man needs to do little things with an attitude of confidence and conviction. A woman needs to respond to the things he does in a receptive but not fully convinced manner . . . A man should not get the idea that she is after him." A woman's pursuit is death to male desire, which needs to experience itself as "active" in order to keep the blowtorch hot. Women shouldn't be cold, and Gray carefully avoids the word "passive" in describing our role. Rather, we should remain "receptive," encouraging the man to do his best to impress *us*, and never, ever trying to impress him. Trying to make *him* happy is a total turnoff to him; let him woo you, always.

Later on, he offers this illuminating analogy. Flirting (woman's role) is like shopping; pursuit (men's) is like a job interview. Women must never, ever seem to be applying for sex or love, but only carefully considering what a man has to offer, perhaps looking in the window for a while before examining the quality of the goods closer up. The successful man, playing his appropriate "complementary" role, puts his best foot forward to "sell himself," "advertise" his services, present his résumé (all Gray's metaphors) to this discerning consumer. Given all this commerce in image, with women holding their own desires like a closed hand in a poker game and men trying to figure out (1) what *she* wants and (2) how to exude "confidence and conviction" without tipping over into offensive behavior, no wonder Mitchell lost track of what he was feeling.

the double bind of masculinity

In 1956, psychologist Gregory Bateson formulated the concept of the "double bind." A double bind is any situation in which a person is subject to mutually incompatible instructions, in which they are directed to fulfill two contradictory requirements at the same time. Bateson's emphasis was on the double messages that parents send to children—for example, demanding that they stand on their own but at the same time encouraging dependency. But the concept is socially applicable too. In a previous book, *Unbearable Weight*, I used the concept of the double bind to explore some of the contradictions that face young women today, living at a time when girls are encouraged to compete alongside men, to "be all that they can be," to "just do it," and so on—at the same time as they take care to not lose their femininity in the process. So we get superbly skilled, fearless young female athletes who speak in baby voices and bat their eyes like Ally McBeal.

Clearly, this culture places young boys in a double bind too. We fabulously reward those boys who succeed in our ritual arenas of primitive potency, and humiliate the boy whose sexual aggression quota doesn't match up to those standards. But at the same time, we want male aggression to bow to civilization when a girl says "no" and to be transformed into tender passion when she says "yes." The fact that these contradictory directives put a *real* person in a difficult (if not impossible) double bind gets masked by the fact that we've created numerous *fictional* heroes who successfully embody both requirements, who have the sexual charisma of an untamed beast and are unbeatable in battle, but are intelligent, erudite, and gentle with women.

Edgar Rice Burroughs's Tarzan is an exemplar. The genetic offspring of British aristocrats, but raised by apes, he combines (as historian Gail Bederman puts it) "the ultimate in Anglo-Saxon manliness with the most primal masculinity . . . violent yet chivalrous; moral yet passionate . . . [and] with a superb body." These qualities make their appearance in every version of the Tarzan story that I've seen, including the recent *George of the Jungle* (Brendan Fraser had the most superb body yet). In a similar vein, Mel Gibson's Braveheart is a fluent linguist, educated in Latin, well-traveled, and a tender, sweet lover. None of that gets in the way, however, when he's called upon to lead an animal-house army of howling Scotsmen against their cruel (and effete) British oppressors. The heroes of *Pulp Fiction* are able to discourse at one moment on abstract intellectual conundrums of philosophy and religion (not to mention cultural variants of the "Big Mac") and in the next, without raising their pulses, blast away at a roomful of human beings.

The mixed messages about manliness get directed at girls as well as boys—and catch them at a very young age. In Disney's most recent version of *Beauty and the Beast*, the beast-hero not only looks and acts quite a bit like a man, but is the keeper of the flame of civilization. True, he must learn to control his anger and to have good table manners. But he owns a high-ceilinged library full of books, and is continually contrasted in the film with the provincial, arrogant, ferociously anti-intellectual hunter Gaston, whom Belle describes at one point as "primeval." (Gaston's room is decorated with antlers; even his furniture is made of antlers.) Gaston, who is a "normal" man, is represented as the truly uncivilized male of the story, because he has no appreciation for beauty (double meaning intended), while the beast, although full of animal rage to begin with, becomes fully tamed under the spell of his love for Belle.

But at the same time as the (tamed) Disney beast represents the virtues of masculine civility, he is also far more romanticized and sexualized *as* beast than in earlier versions of the tale. He growls impressively, rears up on his hind legs in phallic splendor, and has a wild passionate nature that seems almost the cartoon equivalent of Mel Gibson's William Wallace. His aggression is pure of heart, but, like Wallace's, it's pretty potent when it gets going. And this isn't lost on Beauty. In earlier versions of the story—the original tale and various movies based on it, as well as the popular television show—the main idea of the story is that love requires learning to see beyond externals, as the heroine learns to see past the grotesque exterior of the beast to discover the "real" person within (symbolized by the handsome prince who emerges at the end of the story). But in the Disney version, Beauty doesn't really need to see past the beast's externals, for he's pretty magnificent as he is. In fact, as Elaine Showalter writes, "it's a distinct letdown when the Beast turns into a blue-eyed prince"; I agree, the prince is rather lame compared to the commanding figure of the beast. (For Showalter, even the beast was "a bit too gentle for her tastes"; now there's a lover of the wild man!)

The Disney people recognized the erotic charge of their beast; they delayed making the film for a decade, worried about the sexual overtones of the story, until they finally hit on the solution of having constant chaperones around in the castle, in the form of all those talking teapots and dinnerware. But it was the makers of the film, of course, who endowed the beast with all that sex appeal to begin with. As such, the "learning to see past externals" theme becomes merged with, even supplanted by, a version of the "sleeping beauty" motif, whose main point is not the heroine's spiritual education but her sexual awakening. (That's the way contemporary Freudian commentators like Bruno Bettelheim have interpreted the beauty and beast tale too—as about the young girl's initiation into the power of male sexuality.) It's frightening for her at first ("strange, and a bit alarming," as the Disney heroine sings), but it ultimately allows her to unleash the beast within herself. Women, being ovens by nature, need the spark of the male blowtorch to set their instinctual fires going.

The notion that men's more forceful, hotter sexuality is required to turn on a woman's slow-heating oven—an idea which has a certain amount of charm, no doubt about it—is unfortunately also part of the mythology that teaches men not to take "no" for an answer. For in this version of sexual "complementarity," the woman *requires* for her own awakening that the man see past the veneer of civilization motivating her "no" to the *real*, albeit latent, desires within. Of course it's going to take a while . . . she's a slow cooker, remember. But don't give up; she'll come around. That's the way it happens in the movies, isn't it? There, you'll never see the woman who won't let go depicted as anything other than deranged (as in *Fatal Attraction*) or power-mad (as in *Disclosure*). But numerous romantic comedies reward the persistent male suitor for his efforts—with the girl's affections. Popular culture *admires* the man who won't take no for an answer.

When date rape workshops focus on women's learning to say "no" in a way that will be clear and unequivocal, they don't alter this mythology much either. For these (well-meaning) workshops are clearly still imagining women as the gatekeepers of male desire, tamers of the beast, domesticators, while continuing to view men as primitive animals who, if left to their own devices, would naturally plunder everything in sight. The script still has the male in hot pursuit, the woman with the "stop" and "go" signs. Here, women are getting some mixed messages too. John Gray says we should flirt, to stoke the fires of male desire. But popular college training manuals for the prevention of acquaintance rape seem to recommend otherwise. They caution that "flirting," which suggests a certain indecisiveness, a "wait and see what he does for me" attitude, is exactly what should be avoided: "An important part of assertiveness," one of these manuals reads, "includes clearly indicating what is intended with both words and actions. If a woman says 'no' with her mouth, and 'yes' with her body, she is giving a mixed and inconsistent message. She must know what she wants clearly before she can ask for it."

To slightly adapt the refrain that Maurice Chevalier sings in *Gigi*: "Oh, I'm so gla-ad that I-I'm not young . . . nowadays."

"the male animal" as ideology

Destructive double messages aside, it might seem that science ought to have the last word here about what differences in male and female sexuality are "real" and which are the creations of culture. The findings of evolutionary science should be taken very seriously. I think it's also important to take note that scientific ideas about these matters have changed over time, and in not entirely disinterested or objective ways.

It might come as a surprise to contemporary men and women, but throughout the ancient and medieval periods of Western history, it was women, not men, whom science regarded as the hot-blooded, sexually insatiable sex. Women were believed to have a greater sexual appetite than men and to be less capable of controlling their urges. But these qualities carried meanings different from dominant contemporary notions that equate hair-trigger sexuality with virility. Both Greek and early Christian cultures valued sexual self-control highly (although for different reasons) and viewed domination by one's passions as a sign of weakness, *not* manliness. The truly virile man (or, within Christian tradition, the chaste and godly man) was the master of his own body. Being more at the mercy of our sexual urges was thus among the traits that, ancient science argued, were a sign of women's inferiority to men.

Women's lesser control over our sexual impulses was connected to what was believed to be our greater "passivity"—another feature of our natural inferiority. We've already seen, in an earlier chapter, how approval of male "activity" and revulsion with the man who puts himself in a passive, "feminine" position, affected Greek ideas about sex between men (and still inform some homosexual cultures). The same set of values (male = active = admirable; female = passive = inferior) also were embedded in ideas about reproduction, and demonstrate how profoundly gender ideology is capable of shaping even how "hard" scientists interpret their observations. According to Aristotle, the conception of a living being involves the vitalization of the purely material contribution of the female by the "effective, active" element— the male sperm:

> . . . [T]here must be that which generates and that from which it generates, even if these be one, still they must be distinct in form and their essence must be different . . . If, then, the male stands for the effective and active, and the female, considered as female, for the passive, it follows that what the female would contribute would not be semen but material for the semen to work upon. This is just what we find to be the case, for the catamenia (menstrual material) have in their nature an affinity to the primitive matter.

In other words, what the female contributes to conception is mere glop. The forces that will shape that glop into a distinctive human being are all contributed by the male. This view of things was so powerful that when Leeuwenhoek in 1677 first examined sperm under the newly invented microscope, he saw tiny "animalcules" in it—the "form" of the future being, to be stamped onto the shapeless dough of the menstrual matter. We no longer believe in this cookie-cutter version of reproduction, in which female stuff just lies there, waiting to be penetrated and molded by male form. But if you think we've gone beyond mistaken notions about males as "active" and females as "receptive" in the act of conception, consider this account from Alan Guttmacher's drugstore guide *Pregnancy, Birth and Family Planning*:

Some of the sperm swim straight up the one-inch, mucus-filled canal with almost purposeful success, while others bog down on the way, getting hopelessly stranded in tissue bays and coves. A small proportion of the total number ejaculated eventually reach the cavity of the uterus and begin their upward two-inch excursion through its length . . . The undaunted ones, those not stranded in this veritable everglade, reach the openings of the two fallopian tubes . . . The one sperm that achieves its destiny has won against gigantic odds, several hundred million to one . . . No one knows just what selective forces are responsible for the victory. Perhaps the winner had the strongest constitution; perhaps it was the swiftest swimmer of all the contestants entered in the race . . . If ovulation occurred within several minutes to twenty-four hours before the sperm's journey ends, the ovum will be in the tube, awaiting fertilization; if ovulation took place more than twenty-four hours before insemination, the egg cell will already have begun to deteriorate and fragment, rendering it incapable of being fertilized by the time the spermatozoon reaches it. On the other hand, if ovulation has not yet occurred, but takes place within two or three days after intercourse, living spermatozoa will be cruising at the tubal site . . .

Guttmacher's drama of driven, supercharged sperm racing to be the first to penetrate a waiting egg is duplicated in the opening credits of *Look Who's Talking*, which depict the perilous journey of "the undaunted ones" to the tune of "I Get Around," and has the successful sperm (the one "who achieves his destiny") provide a running commentary on his progress. And there, at the end of his journey is the giant beach ball of an egg, languorously bobbing, killing time, awaiting the victor's arrival. But this picture is totally inaccurate as a depiction of a typical act of conception. On most occasions, when fertilization occurs it is actually the *egg* that travels to rendezvous with sperm that have been lolling around, for as much as three days, waiting for *her* to show up. What Guttmacher and *Look Who's Talking* present as their model is an act of intercourse at ovulation—by no means representative of the majority of fertilizations. Yet it has somehow become our paradigm. Could it be that it goes against the grain to imagine males (or their sperm surrogates) passively hanging around, waiting for Ms. Egg to "call"? (To be fair, Guttmacher does allow that sometimes the sperm will be waiting for the egg. But, significantly, he doesn't describe them as "waiting"; rather, they are "cruising," like guys driving their cars down the street, looking for chicks.)

During the Renaissance and Enlightenment, ideas about sexuality were focused less on men and women and more on race, as the group of people who now came to be seen by Western science as having a naturally hotter, more "primitive" sexuality were the "uncivilized" people of non-European descent: African, West Indian, and Native American Indian (both male and female). As Bryan Edward, a Jamaican planter and English politician, wrote in 1793:

> The Negroes in the West Indies, both men and women, would consider it to be the great exertion of tyranny, and the most cruel of all hardships, to be compelled to confine themselves to a single connection with the other sex. Their passion is mere animal desire, implanted by the great Author of all things for the preservation of the species. This the Negroes, without a doubt, possess in common with the rest of the animal creation, and they indulge it, as inclination prompts, in an almost promiscuous intercourse with the other sex.

The notion that the passion of the Negroes is "mere animal desire" was confirmed by pre-evolutionary (and, later, evolutionary) science, which plotted degrees of similarity and difference between the apes and various racial groups; drawings of Africans, the appropriate features exaggerated, were placed closest to those of apes, while sketches of fine-featured Europeans with noble expressions on their faces occupied the other end of the spectrum. Artists depicted "historical" matings between Africans and orangutans. These ideas that the African man (and woman) is closer to purely instinctual, animal nature still survive today, not only in overt racial stereotypes but also in more subtle ways. Actor Laurence Fishburne was recently depicted in *Vanity Fair* in a series of photographs posing him in menacing, lionlike postures, with captions like: "Laurence has natural grace, but he's not afraid to be savage" and "He's like a dangerous, magnificent beast."

How did we get from the identification of animal sexuality with non-Europeans of both sexes to the current notion that *men*—of all races—are hotter-blooded and more promiscuous in nature than women? It wasn't discoveries in evolutionary science that led the way, but changes in perceptions of what constituted the ideal man in an industrial, increasingly competitive society. By the end of the nineteenth century, Europeans began rethinking their attitudes toward the primitive "savage," not out of any sense of morality or political correctness, but because the primitive savage was beginning to be seen as having something the European gentleman lacked and needed. As we've seen in an earlier chapter, it was a time when the prestige of older notions of manliness was eroding, since those notions revolved around qualities no longer very useful to success in a market-driven economy. Being a "civilized gentleman" didn't get you very far in the competitive jungle of the marketplace. At the same time, "civilization" itself was increasingly being viewed as a source of human "discontent" (as Freud put it) responsible for numerous new nervous disorders seen as being caused by the stresses and strain of modern industrial life.

One of the most interesting expressions of this new mistrust of the suppression of instinct by civilization is found in the French nineteenth-century literature on rabies, which imagined the disease as an eruption of the beast through the suffocating veneer of civilization, turning the human sufferer into an animal, wild, uncontrollable, and dangerous, prone to biting and tearing apart everything in sight. Remarkably but consistently, rabies in dogs too was viewed as the result of hypercivilization and sexual repression! It was believed that dogs in the wild rarely developed rabies, that it was produced by the "unwholesome conditions of domesticity" where dogs are forced into "unnatural chastity by their mistresses."

In this context of growing concern and anxiety about the repressive effects of civilization and its "softening" of men (and dogs), fantasies of recovering an unspoiled, primitive masculinity began to emerge, and with them, a "flood of animal metaphors" poured forth to animate a new conception of masculinity. White men drew on the images and ideology of the savage Other to help them articulate this emerging construction of "passionate manhood" (as historian Anthony Rotundo calls it). The depiction of historical scenes of European rape and pillage—a motif that was formerly most prominent only in racist representations of the black male rapist or marauding Indian—became especially fashionable. (On page 250 is an unabashed, very politically incorrect—by today's standards—1932 version ["Nature in the Raw is seldom MILD"]). Africans and other "primitive" peoples were also playacted in clubs and lodges. Bourgeois, white, American men begin to speak of themselves in terms of "animal instincts" and "animal energy." They began to believe that "this nature was their male birthright and that it demanded expression."

In this way, arguably, the notion that men are passionate beasts by

Half a century after "Nature in the Raw," the appeal of the untamed male animal persists in contemporary notions of masculinity, from *Esquire*'s only slightly tongue-in-cheek celebration of "Wild Men" to Disney's lovable yet phallic beast (*Photofest*)

nature, who cannot and should not be expected to control themselves, gained cultural cachet—a cachet that has created a special sort of double bind for black men. Within racist mythology, being more instinctual means being a mere animal, a brute. But to the extent that being an untamed animal has become part of the definition of being a potent man, then being associated with sexual instinct and animal magnetism gives one the gender edge. It's no wonder, then, that black artists and athletes have sometimes exploited the stereotypes that endow them with greater potency and "animal" sexuality. Particularly when one has been historically deprived of most of the social privileges of masculinity, it would be hard to relinquish the one advantage one has. Unfortunately, the advantage comes at a high price, for that old racial mythology is still there, ready to be activated. When a white boy acts like a thug, he proves he's not a sissy to the other white guys in his group; when a black boy engages in the same behavior, the same white boys may view it as proof that he's a jungle brute after all.

Today, with many men feeling that women—particularly feminists—have been pushing them around for a couple of decades, the idea of a return to manhood "in the raw" has a refreshed, contemporary appeal. The "Return of the Alpha Male" literature and a good deal of the mythopoetic men's movement seem clearly to be "backlash" reclamations of manhood, which, like some of their Victorian counterparts, view women as responsible for having tamed the beast in man. Other Victorian conceptions, as we've seen, put the chief blame on civilization—and plenty of contemporary ads exploit that notion too, urging escape into the wild (in a luxury four-wheel-drive car). In neither case, however, is the ideal *really* for men to revert to the jungle (except perhaps on a weekend jaunt). For at the same time as we romanticize the wild man and scorn the wimp, we also—like the Greeks—still place great stock in masculine self-control. It's just that we expect it to have a potent, aggressive edge.

Such notions, and the ambivalence about "civilization" and "nature" that underlies them, are vividly illustrated in Mike Nichols's recent contemporary reworking of the werewolf story, in the movie *Wolf*. This film plays with the same idea which was so prominent in the Victorian era—that hypercivilization has made man soft, ineffective, and deadened and that his revitalization requires reconnection with his "animal" nature. Jack Nicholson's character, Will Randall, is a book editor at—oh, dear—"McLeash" House, where he performs his job with "taste, individuality and civility" and is given, at parties, to bemoaning the death of art and the triumph of TV talk shows. He's also a wonderfully "nice" person, as his boss Raymond Alden tells him, explaining why he's replacing him with his unscrupulous, ambitious assistant (who is also having an affair with Randall's wife). "What are you, the last civilized man?" the boss's daughter (Michelle Pfeiffer, wearing a buckskin jacket, by the way) cracks when he apologizes for finishing her drink, spelling it out for those of us who haven't gotten it yet. Randall is civilized, all right. But Nichols can't decide whether or not that's a good thing.

On the one hand, his niceness and love of art and gentleness are favorably contrasted with the ruthless crudeness of his boss and assistant, and the heroine explains that she loves him because he's a "good man" and that's "very exotic" to her. On the other hand, we are encouraged to see him as wimpy. He accepts being fired passively and uncomplainingly; his feminist wife (Kate Nelligan) nags and scolds him. At the beginning of the movie, the night he gets bitten by the wolf, he has just returned from closing a deal with an author; when asked by his wife how he got the author to sign the contract, he says he did it "the old-fashioned way; I begged." But being bitten changes all that for him. He gets in touch with his sexuality and aggression, makes love "like an animal" (as his wife tells him), begins to crave raw meat, and starts stalking game at night. And he gets even in primitive but effective ways with those who cross him. (For example, in a Bernhard Goetz white-man retribution moment, he bites off the fingers of a black gang member who tries to rob him in the park. In another scene, which must have delighted Camille Paglia, he pisses on the shoes of his assistant: "Just marking my territory and you got in the way," he says.)

All these, arguably, are in keeping with traditional werewolf stuff, albeit modernized. But becoming a wolf also makes Will more effective in the world of business—he sees and hears acutely, and uses this to his advantage—and most strikingly, his new wolfness makes him adept at the very kinds of ruthless machinations that the movie had portrayed so negatively earlier and juxtaposed against Will's civilized "niceness." He engineers a brilliant yet deceptive scheme to get his job back, and earns the respect of all around him. "You're my god," says his faithful editorial assistant (David Hyde Pierce) when he discovers that Will's scheme to get his job back is a bluff, based on a lie. And when the scheme succeeds, his boss tells him that if he'd known Will was "this ruthless, I'd never have fired you in the first place."

Here, the idea that real manliness (and sexual vitality and zest for life) is to be found outside man-made culture is merged with the idea of the workplace as the man-made jungle where a man might realize himself, if he's the right sort of animal. Will's transformation gives him the right stuff, and in the process he becomes something of an incoherency. He begins to speak like a character from a whole other movie. "The worm has turned and is packing an Uzi," he declares to his secretary. The bite of the wolf has turned Jack Nicholson into Arnold Schwarzenegger! Wolf man as terminator? Somehow I think the metaphors are getting mixed. In the end, the movie does suggest a final exit, with Michelle Pfeiffer, into the real jungle, where they will live fully as animals. But until then, the wolf man is allowed to have it all—spontaneous and playful animal sex with Michelle Pfeiffer, carnivorous romps in the woods, aggressive vengeance against his enemies—and more business smarts than Malcolm Forbes.

examining the science of sexual difference

I suspect that John Gray has given very little thought to the history of his notions about the sexual differences between men and women, or the ideologies with which they have been entangled. As far as more scientific versions go, scientists probably don't think of them as "notions" at all but as "findings." Of course, they can be both, and they *are* both. Science has much to teach us, but only if we subject it to rigorous scrutiny—and unfortunately *Time* and *Newsweek* rarely do that when reporting about "new discoveries" concerning the "science" of sexual difference.

The studies of the social scientists are the most problematic. It's an unfortunate limitation on the science of sexual difference that no sociological data, no psychological experiment (that anyone has come up with yet, in any case) has indicated how to clearly sort out "culture" and "nature" in human subjects. The interactions of the two are too complex, too subtle, too intertwined. At most, we can point to suggestive data. But even with that qualification, the leap from observations of behavior to claims about "nature" is fraught with peril. "An abundance of physical energy and the urge to conquer—these are normal male characteristics, and in an earlier age they were good things, even essential to survival," writes Barbara Kantrowitz in *Newsweek*, in an article called "Boys Will Be Boys." The quick slide from "abundance of physical energy" to "urge to conquer" is barely noticed by the casual reader, although the notion that young boys have a natural "urge to conquer" is much more controversial than the fact that they seem to run around and knock over furniture more than girls.

Kantrowitz's understanding of our evolutionary heritage seems to be the crude—and mistaken—notion that we've inherited certain predilections, carried on certain genes, in some pristine, eternal, and incontrovertible form. That's not the way evolution works; in fact it's exactly the opposite of the way evolution works. "In countless instances," writes biological anthropologist Richard Wrangham, "biologists have watched animals change their behavior to suit their purposes. Indeed, the whole logic of evolution would indicate that animals use their intelligence to serve evolutionarily appropriate goals. Why otherwise would problem solving and learning (and the variable behaviors that these abilities create) ever have evolved? Complex animals have complex mental and emotional systems underlying their behavior. These systems have evolved, and they are subject in turn to genetic variation."

The human brain is not fully developed at birth. In the first two years of life, it increases enormously in size and develops countless synapses that it did not have at birth. During this period, as William Pollack points out, the brain is "pliable and plastic" and tremendously open to learning from experience. It's "wired to accommodate developmental interactions that further shape the nervous system after birth." So it's in our "nature" *not* to have our script given to us at birth. That makes good evolutionary sense, given the tremendous environmental diversity and challenges that human primates, dispersed all over the globe, have had to deal with. Even seemingly "pure" biological markers—like testosterone—do not unambiguously indicate any one, clear set of qualities that are "natural" to men. Testosterone levels are measurable. But just what is indicated about contemporary men when they rise and fall may not be—probably is not—the same thing that was indicated when they rose and fell among our early ancestors. And there's tremendous *individual* variety too in the way testosterone functions among men—that too is part of our "nature."

Testosterone levels rise in response to sexual excitation (there are anecdotal reports from transsexuals receiving male hormones of an "almost unbearable" increase in sexual drive). But testosterone responds in men to other factors besides sexual excitement. The levels of professional athletes rise steeply in anticipation of competition, and go up even more following a victory. How to interpret this? The pre-game rises may look like an "urge to conquer." But what about those hormonal victory celebrations? Even harder to interpret are studies that suggest that testosterone levels in men also rise in response to marital difficulties. Moreover, as William Pollack points out:

"Testosterone can have a variety of different effects on a boy's behavior. A high testosterone level in one boy may enable him to play a chess match with great intensity and alertness; in another, it may give him the energy and concentration to make complicated arrangements for a political rally. In a third case, it may contribute to his involvement in a brutal fistfight. . . . In older men the effects of testosterone vary from man to man. Research has also shown that when older men are given supplemental testosterone they become calmer and less aggressive than before."

A great deal of rigor is gained by social science studies of sexual difference when the populations they study have some ethnic, racial, and historical breadth; when patterns manifest themselves over and over, despite the diversity in groups and historical eras, one might suspect that some kind of biological pressure—which then, of course, would still need to be interpreted—is being exerted. Yet sociologists and psychologists intent on proving hard-wired differences in sexual temperament between men and women continue to do their surveys and measure the responses of subjects with little attention to possible cultural or historical variables Many of the psychological studies of men and women's different sexual responses to various sorts of sexual stimuli, for example, are blithely ethnocentric in their choice of subject populations. They rarely extend their experiments to subjects outside the United States, and make virtually no accommodation to possible differences among ethnic and racial groups. They pay little attention to the varied cultural meanings attached to nakedness, or certain styles of dress, or particular gestures.

"In humans," popular science writer Deborah Blum writes, "the male system sometimes seems so jittery with sexual readiness that just about anything—high-heeled shoes, a smile, a friendly conversation— will produce a sexual response." Is she talking about *all* male humans here, or just those she has observed hanging around her hometown bar at happy hour? I wouldn't have any problem with Blum's observation, unspecific as it is, if it weren't offered alongside various other studies in an attempt to provide massive, cumulative evidence of hard-wired differences between the sexual appetites and sexual "readiness" of men and women. These studies run the gamut from comparisons between humans and birds to those outdated Kinsey surveys of men's and women's different responses to pictures of naked bodies.

All piled together, it can look pretty convincing. But each block in the edifice, examined carefully, doesn't stand very sturdily on its own. The evidence acquired by looking to other animals is certainly not de- finitive, since animals display a wide range of variation in their mating and "dating" patterns. (The females of some species of monkey, for example, are "promiscuous" and highly sexually excitable.) The sur- veys of humans all appear to be of American populations, and par- ticularly when it comes to sexual attitudes, that can yield highly misleading results, especially when the racial and ethnic mix is not diverse. (I have no idea whether that's the case with the studies Blum cites, since she never mentions race or ethnicity as potentially mean- ingful variables.) Although white Americans tend to think of them- selves as the norm in all matters, when it comes to sexuality they are in fact rather odd birds. As is more obvious in the late 1990s than ever before, a salacious, guilt-ridden, adolescent obsession with sex con- tinually leaks out of every national pore. That's surely a relevant cul- tural consideration to add to those observations of American men's jittery responses to high heels!

Many of the conclusions Blum draws from the studies she discusses lack historical perspective too. As I argued in an earlier chapter, women's seeming sexual indifference to visual stimuli as reported in earlier studies appears to be changing, as women are given greater cultural permission—and more opportunities—to be open voyeurs.

Blum's book (called *Sex on the Brain: The Biological Differences Between Men and Women*) was published in 1997; by then, the cultural landscape had already changed dramatically, suggesting at least the possibility that new studies might yield some new findings. Yet Blum cites that old Kinsey survey that shows women don't respond to pictures of naked men in the same way that men respond to pictures of naked women as though it were a window onto our differing biological makeup. (She swiftly dispenses with the possibility that culture may play the decisive role by pointing out that society has tried to repress men's interest in dirty pictures, but it has failed. I think the reader of this book will immediately see the flaw in this reasoning. Only a die-hard Skinnerian would believe that punishing the little boy who is found in the bathroom with *Playboy* is going to make those pictures of naked women less titillating to him. Probably the punishment will have just the opposite effect. Having few naked pictures to *get* titillated by—as has mostly been the case for heterosexual women until recently—is a whole other matter.)

I'm not saying, I want to emphasize, that there *aren't* hard-wired differences in sexual temperament between men and women. I'm just urging that we exercise caution before we make leaps from observations of behaviors that are, inevitably, soaked through with culture—as all human behaviors are—to declarations about "sex on the brain." If I wasn't a skeptic about such declarations before, I certainly became one when I read in the newspaper, just last month, that Scottish psychologists had discovered (in a study of 92 men and women in Japan, South Africa, and Scotland) that both sexes find a "slightly feminized" male face (such as Johnny Depp's or Leonardo DiCaprio's) to be more attractive than the "high-testosterone" looks of a George Clooney or Arnold Schwarzenegger. The high-testosterone looks were associated with coldness, dominance, and dishonesty; the more "feminized" face with warmth, fidelity, cooperation.

For the purposes of my point I'll pass over the absurdity of putting Arnold Schwarzenegger and George Clooney in the same category, and point out instead that two years ago sociologist Alan Mazur found exactly the opposite: that "consistently across cultures . . . the man who turns heads is not necessarily beautiful but is powerful and inspires fear and respect." Facially, he has "a dominant appearance—a muscular-looking face, with a strong chin and brow and good skeletal structure." Schwarzenegger was again cited as an example—this time of the preferred kind of man, with physical characteristics that are associated, Mazur claims, not only with social leadership (Lenin, Ike, Patton, and Colin Powell were other examples of guys who had the Look) but with "skill as lotharios." (Evidence? They tended to have sex at an earlier age than others.)

What's funniest about all this is not the diametrically opposed findings of the two studies—which is humorous only when you consider the absolute authority and conviction with which each struts its stuff—but what those Scottish psychologists make of their study. One might assume that if either of these studies suggests that beauty standards are still driven by traditional evolutionary concerns, it's Mazur's. Those high-testosterone guys, after all, are the ones who are likely to be physically stronger, healthier, better hunters, better protectors. Contrastingly, the sudden preference for warm, cooperative, "feminine" men (if indeed there is such a preference—I'm not convinced)* would seem most likely reflective of recent social and cultural shifts—for example, the changing needs of families in which mothers and fathers both work full-time, the cultural loss of authority of the Superman figure, the influence of gay and queer aesthetics on fashion. And so on. It makes the most sense, doesn't it?

Not to the Scottish psychologists. No, what their study proves, they claim, is that "beauty is not so much in the eye of the beholder; it is hard-wired in the brain." How do they get there? It is standard pop evolutionary wisdom nowadays, you will recall, that women have a tremendous amount invested, evolutionarily speaking, in each egg and its safe growth to adulthood. And while the testosterone king may be better able to protect offspring from predators and bring home food (qualities that once were essential to long-term survival of young), he's also more likely to stray. Here, David Perrett—one of the Scottish researchers—cites studies of U.S. military men, which show that those with high levels of testosterone have "more broken marriages, more violent behavior, and less sexual satisfaction." More "feminine" men, on the other hand, "give the appearance they'll be more likely to stick around in the morning and help raise the children."

---

*The study, remember, was of only 92 people, representing only three cultural systems. And it didn't ask anything about bodies, only faces.

The question arises, of course, as to why women have only discovered the beauty of the feminine male face just *now*, if it is "hardwired" to suggest the kind of man who would serve their evolutionary purposes so wonderfully. Perrett might answer that it is only now—in the final decade of the twentieth century, a nanosecond in evolutionary terms!—that an emotional willingness to share child care has come to be more evolutionarily desirable than the physical ability to vanquish a marauding neighbor. If so, how did that preference get to be "hardwired" so quickly? Perhaps what Perrett considers "hard-wired" here is the female investment in her eggs, which expresses itself in different ways under changed environmental, social, economic conditions? Okay, let's say we accept that. It's a far, far cry from the conclusion that beauty standards are hard-wired. And why such an overriding need to square such culturally suggestive evidence with an evolutionary hypothesis? Could it be that Perrett is a priori committed—whatever the empirical data—to the notion that *any* new beauty preference has its ultimate roots in the interests of reproductive biology?

If so, I wonder how he would explain many men's current preference for estrogen-poor, skinny women? The real evolutionary scientists know better. They remind us, as Jared Diamond does, that "perhaps our greatest distinction as a species is our capacity, unique among animals, to make counter-evolutionary choices." This doesn't mean that our evolutionary past has left no surviving imprints and influences on our behavior. In fact, one of the most fascinating studies on the "male animal" that I've read thus far argues that it has, drawing provocative comparisons between patterns of male aggression between humans and our closest primate relatives, the chimpanzees.

Richard Wrangham and Dale Peterson's study begins with some chilling facts. Of four thousand mammals and ten million or more animal species, it's only among humans and chimpanzees that males make lethal raids into neighboring communities in search of vulnerable enemies (within their own species) to attack and kill. Battery of females, which is in general very rare in the animal world, is much more common among humans and chimps. Rape occurs much more often among the great apes—orangutans in particular, but often among chimps (and humans too, of course)—than among other animals. Human males and some of their closest primate relatives (not all, as becomes important to the story later) appear to have a capacity for

violent gang-bonding, territorialism, and terrorism against females
that is unique among animals. Wrangham and Peterson call these
species "demonic males."

The thesis, on the face of it, seems so male-unfriendly that if a
female biologist had come up with it—not to mention that label
"demonic males"!—I suspect it might be dismissed as feminist male-
bashing. Wrangham's ultimate argument is indeed feminist—"evolutionary feminism" is what he calls it—but its point is not to bash
males. Instead, he unearths the *conditions* that produce lethal male violence. To do this, he compares the "demonic" apes to the most gentle
of apes—the bonobos. The bonobos have been most fascinating to researchers because of their sexual habits, which go way beyond sex for
purposes of procreation. The bonobos have sex—homosexual as well
as heterosexual—in as many positions as they can dream up, beginning
long before the onset of puberty—not only to have babies but as a way
of making friends, of making up after a fight, of comforting each other,
of playing. Part of the reason for this, researchers speculate, is the fact
that ovulation among the bonobos—as among humans—is relatively
concealed. Thus, sexual "anxiety" centered on reproduction isn't particularly productive, since a male never knows when he's going to hit
the mark anyway. So why not just kick back and have some fun?

In addition to their idyllic sexual lives, the bonobos of neighboring
communities show no signs of wanting to kill each other, and often
treat each other with the kind of friendliness they show within communities. There's less male competition for rank or mates, and there
are no reports of any battering or rape of females. When males do—
very rarely—exhibit aggression against females, they are liable to be
driven off by a gang of females. Females are also able to break up incidents of aggression between males. The sexes are "co-dominant"—
that is to say, the most powerful females have equal status with the
most powerful males. Also striking is the closeness and mutual support
that exist between bonobo mothers and sons: a young male's most important relationship is probably with his mother. The reason, Wrangham and Peterson suggest, is the extraordinarily strong community
bonds that exist among bonobo females. These female bonds are much
stronger, in fact, than the bonds among bonobo males, who rarely co-
operate with each other in either defense or attack. It stands to reason

that if you're threatened, you want to have the females on your side. They have the power of numbers, and solidarity. "Even the highest-ranking male," the authors report, "can be defeated when females gang up on him." But they rarely have to.

The reasons why the bonobos have evolved in this way, while chimps and humans have such strong tendencies toward "demonic" male bonding are complex, involving the different food supplies and other features of the environments in which the different species have struggled to survive. What's most interesting for my purposes here are the social structures that have developed that hold the potential for male aggression and violence in check. They are: (1) absence of strong male gang bonding, (2) female bonds of solidarity, and (3, and I believe most important) resulting conditions of sexual equality and intimacy among males and females. I read *Our Guys*, Bernard Lefkowitz's gripping account of the Glen Ridge rape, a bit after I discovered *Demonic Males*. As I was reading *Our Guys*, those bonobos kept coming into my head, but I couldn't precisely figure out why until I went back to reread *Demonic Males*. Then it hit me like the proverbial ton of bricks. The Glen Ridge community, besides being a paradigm case of demonic male behavior, was almost the mirror image of bonobo life. Glen Ridge was characterized by:

(1): Intense bonds among the athletes, frequently channeled into either hostility against other men of the "wrong sort" or gang abuse of women, including vandalism of property with little point other than to push one's weight around, "voyeuring," and ultimately gang rape.

(2) Virtually no solidarity among the community of women. The girls baked cakes for the athletes, decorated their lockers, protected and in general doted on them. When other girls were sexually abused, they didn't come to their aid but instead blamed the girls for their own stupidity, gullibility, or trashiness.

(3) Conditions of equality and intimacy were far from existing among males and females. Girls were viewed by boys as simply being there to serve male needs, either as "Little Mothers" who helped with homework and baked cookies or as sex toys for providing sexual release. They almost never had sexual intercourse with these girls, preferring hand jobs and blow jobs—sex in which girls, as Lefkowitz describes it, "performed at their bidding. The guys were the foremen

supervising their work crew." The "Little Mothers" were never allowed to get close to the boys either. "The guys were really close only with each other," recalled one.

The schools didn't respect girls either, *or* treat them with the same leniency they showed the boys. They brushed it off when the boys engaged in harassment, exhibitionism, and, ultimately, rape—describing the charge as "alleged sexual misconduct" and cautioning the student body to "not be judgmental" about the young men who had been arrested. In contrast, a young woman who got drunk in public was severely penalized, removed from her position as president of two prestigious school organizations. "She lost everything. She was extremely intelligent and she screwed up once," policewoman Sheila Byron recalled, "whereas these boys—a two-day slap on the wrist. Doesn't seem fair, does it?"

Not all American communities, of course, are like Glen Ridge was in 1989—although many may have important features in common with it. Not all men, certainly, behave with women like the Glen Ridge athletes, or Mike Tyson. As a species, we may have the *capacity* to produce "demonic males," but we certainly have the capacity to produce decidedly nondemonic males too—and do so all the time. So the question is not: are all men sexual predators and aggressors? The answer to that one is clearly: no. The as yet undecided question is how we account for and understand male sexual predation and violence when it does occur.

One way we explain it, as seems the (unacknowledged and unconfronted) implication of many of the pop theories circulating today, is as the unfortunate but unavoidable consequence of a high-testosterone, genetic inheritance that may no longer be as socially or evolutionarily beneficial as it once was, but that we just gotta learn to live with. This is to view men's "inheritance" as equivalent to my little Jack Russell's inability to stop trying to bury that bone in places where it simply can't be buried—like a leather sofa. The poor little fellow just keeps pushing his nose over that smooth surface as though it is a pile of dirt that can be dislodged. Finally, I just have to take the bone away from him in pity. I adore my little dog. But as smart as I think he is—

for a dog—human brains have evolved to be much more adaptable to their environment than his has.

A second way to look at male violence and aggression is as entirely a cultural overlay. Those who try to make this argument produce evidence of cultures where men are more like bonobos than they are like demonic apes. I haven't concentrated on those arguments here, because I grant their evidence of cultural diversity, but do not view it as "answering" the question of how to account for male violence where it does occur. I have to admit, too, that when social construction is taken to an extreme, as it is by some, it goes against my grain as much as simpleminded biological determinism. "The male/female dichotomy has no intrinsic biological reality," states the introduction to a recent collection of articles on masculinity. The authors aren't just talking about specific gender traits (proposed cognitive differences, for example), but the very fact of sexual difference itself, which is viewed by many academic theorists as the product of biologistic "discourse." In other words, human beings are responsible for the very division of creatures *into* "males" and "females." It's just a heterosexist "trope," a product of language. What about the obvious differences—in genital structure, for example? *We* are the ones who have made those count as the dividing line between two sexes. In my opinion, this goes way too far in the direction of an arrogant homocentrism (an old impulse in new "linguistic" form) that lifts human beings out of the evolutionary picture entirely.

A final approach, which Wrangham and Peterson advocate, and which represents my own view, is to recognize that our biological inheritance itself allows for a great deal of creativity and intelligence both in adapting to new environments and in adapting environments to new goals. This means that evolutionary biology would *expect* male and female "nature" to be shaped by the diversity of human environments in interaction with hormones, physical morphology, and so on. It means that there are always repertoires of possibilities, never single paths. It also means that we need to study the contexts and conditions of the paths we *have* taken—to be discovered through comparisons with other primates, examination of environmental pressures, and so on—in order to intelligently assess the paths we haven't taken that might still be open to us.

The specific lesson I take from the demonic male/Glen Ridge comparison is not that females ought to take over culture and start pummeling roughneck little boys into gentle, sweet, bonobo shape. The lesson I take is that acting as if boys and girls are from the same planet—and treating them that way—may be our best safeguard against the kind of group violence and aggression that the Glen Ridge athletes engaged in. "Women are from Venus and men are from Mars" and all those other current encouragements for men to regard women as an alien species and themselves as testosterone-driven beasts that only their "brothers" can truly understand may be not only self-fulfilling but harmful—to men as well as women. William Pollack has detailed the psychological costs to boys of insisting that they detach prematurely from their mothers and become tough, exclusively male-identified little macho men. And when males grow up isolated and bonded within a separate male "culture" dedicated to "letting boys be boys," scrutinizing females—those odd creatures—with bafflement and anxiety (and sometimes scorn), bad things can happen to women.

This doesn't mean ignoring the specific needs, abilities, challenges, and delights of little boys—*or* men. We can admire athletes without turning them into top-ranking males of some primal tribe. We can find beauty in the muscular male body without celebrating "The Return of the Alpha Male." And we can let little boys play rough-and-tumble, compete with ferocity, even knock each other to the ground, without acknowledging some "urge to conquer" implanted in their genes.

the sexual harasser is a bully, not a sex fiend

beyond grabs and gropes

During the 1990s, I've watched with dismay as sexual harassment has become equated with crude sexual overtures, a confusion that reached an apotheosis with Paula Jones's charges of sexual harassment against President William Jefferson Clinton. According to Jones, Clinton arranged to have state trooper Danny Ferguson bring Jones up to his suite in the Excelsior Hotel. There, Clinton exposed himself to her and asked her to "kiss it." This overture, Jones's lawyers claimed, was "so outrageous in character, and extreme in degree, as to go beyond all possible bounds of decency, and to be regarded as atrocious and utterly intolerable in a civilized society." There was talk of specialists being brought in to testify about the shape and slope of the President's penis, which Jones claimed was distinctive.

When Judge Susan Webber Wright threw Paula Jones's charges out of court in April of 1998, arguing that even if Clinton had done everything Jones had claimed, it did not constitute sexual harassment, I was encouraged. I thought that perhaps some important and abandoned questions, which had first emerged in the public discussions during the 1991 Clarence Thomas hearings, would be revived. What's the difference between a crude pass and sexual intimidation? When and how does a highly sexualized workplace or classroom-jokes, sexist comments, calendar art—become a "hostile" and discriminatory environment for employees and students? If sexual harassment is so horrible to endure, why do so many women remain silent, stay on their jobs, even behave in a friendly manner with their alleged harassers? Does sexual harassment *have* to involve physical gestures, dirty talk, propositions, or are there forms of harassment that have nothing to do with sex at all? Do *all* unwelcome sexual overtures compromise equality in the workplace?

Judge Wright answered that last question with a resounding "*no.*" Perhaps, I thought at the time, that wise answer would be an occasion for us to rescue the distinctions that had been buried by the grabs-and-gropes conception of sexual harassment that had developed since 1991 and the Clarence Thomas hearings. But renewed clarity and reason weren't to prevail. For at the point at which Judge Wright made her ruling, Ken Starr had already been off and running for some time, solidifying—no, advancing—the debasement of public discourse about harassment. Paula Jones's lawyers had snagged no instances of sexual harassment—only a lot of dirty details—in their fishing expedition into Clinton's sex life. Even Starr himself has never argued that Clinton was a sexual harasser. But in the deluge of outrage over Clinton's behavior with women that Starr instigated, the difference between illicit sexual acts and illegal sexual acts kept getting lost, and the public understanding of harassment plunged even deeper into the toilet than it had been.

Feminists tried gamely to make some distinctions between sexual crimes and evidence of unrestrained but mutual libido. Gloria Steinem, writing in March in *The New York Times*, correctly pointed out that Monica Lewinsky's relationship with President Clinton had "never been called unwelcome, coerced or other than something she sought." Such attempts to promote a more nuanced conversation were pretty well muzzled by the prevailing sound bite on feminists who supported Clinton, which was that we were hypocritical, partisan elitists who wouldn't champion victims with big hair. But even Steinem's response missed the boat in an important way. I wholeheartedly agree with her that our legitimate public interest in Bill Clinton's (or any elected official's) sexual life should only begin when abuse of power is at issue. But for Starr's tenacity, his unwillingness to let go of the opportunity that the Jones case had opened up for him (or, on some accounts, been orchestrated by him), Wright's ruling on the Jones case should have been the end of the story.

Where I differ with Steinem, however, is in her implicit suggestion that whether or not Clinton (or anyone else) is guilty of sexual harassment turns on whether certain sexual acts were welcome or coerced. This still keeps the focus on grabs and gropes at the forefront of our discussion of sexual harassment. Anita Hill, more astutely, reminded us in her *New York Times* opinion piece that although sex makes a hot story, sexual harassment is not really about sex. Yes, sexuality may be one of the *mediums* of harassment. But it is not its essence.

What *is* its essence? It is evoked, although not defined, in the following account of a female welder:

"It's a form of harassment every time I pick up a sledgehammer and that prick laughs at me, you know. It's a form of harassment when a journeyman is supposed to be training me and it's real clear to me that he does not want to give me any information whatsoever. He does not want me to be there at all . . . They put me with this one who is a lunatic . . . he's the one who drilled the hole in my arm . . . It's a form of harassment to me when the working foreman puts me in a dangerous situation and tells me to do something in an improper way and then tells me, Oh, you can't do that! . . . It's a form of harassment to me when they call me honey and I have to tell them every day, Don't call me that, you know, I have a name printed right on my thing . . . Ah, you know, it's all a form of harassment to me. It's not right. They don't treat each other that way. They shouldn't treat me that way."

Judge Wright's ruling made clear that not all unwelcome sexual advances compromise equality in the workplace. It's also true, as the welder's awful story illustrates, that one can compromise equality without saying or doing anything sexual at all. Except for quid pro quo instances—when, for example, promotions or grades are promised in return for sexual favors—most "sexual harassment" is an unfortunate misnomer, perpetuating an even deeper mistake in thinking. "Sex" can refer to two different things: the classification of people into beings that are "male" or "female" or the performance of certain acts. Sexual harassment law has been focused on the latter sense of "sex" when it *ought* to be focused on the former. Its target, in other words, ought to be gender discrimination: behavior and remarks, sexual and otherwise, that deliberately make a point of sexual "difference," making women feel less than fully accepted or competent in the classroom or workplace. It's really no different from racial discrimination, except

the fulcrum of the harassment is gender rather than race. Would anyone question that a white boss who called his black employee "boy"
was harassing him and creating a hostile and discriminatory work
environment? In the welder's account, her co-workers' calling her
"honey"—not a one-time occasion but ongoing, despite her protests—
functions in exactly the same way.

Does this mean that every time a man calls a co-worker "honey,"
he's guilty of harassment? Some of those who hold to the grab-and-
grope conception of harassment might argue so, because they believe
that unwelcome sexual gestures are inherently demeaning and under-
mining. This raises a whole host of problems. For one thing, how does
one know what's "welcome" or not before the fact? (One grim answer,
installed in a few college code books, is that you ask before you do
anything.) For another, what constitutes a "sexual" gesture as opposed to a gesture of, say, affection? It's not always easy to determine
this. The language of racial abuse, in general, is much clearer, more
commonly understood than the language of sexuality. "Boy," spoken
to a grown-up black man, is pretty unambiguously condescending.
"Honey" is not so clear. Different ethnicities, even different personalities, speak different sexual languages, both bodily and verbal. I personally use the term "honey" to communicate sympathy and maternal
concern, sometimes even to people I hardly know. Most men, nowadays, wouldn't dare use the term in a similar fashion. In my opinion,
that's a sad commentary on the formulaic nature of our cultural understanding of sexual harassment.

Once we shift our focus from sexual gestures and words per se to
patterns of unequal or abusive treatment within which sexual gestures
and words may or may *not* play a role, context becomes decisive. I had
an older colleague who continually called me "honey" as he passed me
in the hall. I found it to be demeaning, and I told him so—rather gently, to allow for the possibility that I had misinterpreted, and also to
enable him to change his behavior without feeling attacked. We wound
up having a pretty interesting conversation, at the end of which he expressed gratitude to me; it was the first time, he said, that any woman
had ever talked to him about such issues. He really hadn't known that
I was offended, and he was glad to have found out. Contrast *this* situation to the one described by the welder, within which the use of

"honey" is part of an aggressive, belligerent, provocative pattern of disrespect, *fueled* by the knowledge that she is offended.

From this perspective, a boss's putting a calendar with naked women on the wall of an office is not in itself necessarily harassing; nor is a teacher's putting an arm around a student. When do sexual gestures *become* harassment? Once you move beyond the crude focus on grabs and gropes, things can get complex. It's clear that a boss or teacher who willfully ignores the objections of subordinates—whether to sexist comments, sexual gestures, or any other form of intimidating or denigrating behavior—is engaged in a pattern of conscious harassment. When subordinates don't object, the situation becomes murkier, but doesn't necessarily mean that harassment hasn't occurred. Subordinates are not always—perhaps rarely—in a position to freely reject, object, or offer moral instruction to their superiors. Their jobs and grades may be at stake, but more subtly and profoundly, they may experience themselves as overwhelmed by the sheer cultural power held by their bosses and teachers—their power, that is, to control the operant language, interpretations, "reality" of the situation.

Often, in fact, harassment *begins* when that power is challenged. This was the case with my own experience of harassment, which I've described in earlier pieces, when I was in graduate school, in the days when we didn't have a word for such things. It had begun with a professor's expressing more than professional interest in me over lunch. I wasn't interested and I indicated as much, but I didn't feel offended or compromised. The fact that I was in a class of his didn't concern me, because up to that point I had no reason to believe that what had occurred at lunch would affect his treatment of me as a student. We were more naive in those days, true. But I also do not believe, even today, that emotional complications of this particular nature necessarily compromise teacher/student relations more, say, than unexpressed fantasies or covert resentments. The man had made a pass, I had said no. Big deal.

I began to feel uncomfortable only when my professor, having been turned down, began to sprinkle virtually every conversation we had with references to my gender and my personal life. After I passed my comps, he told me that he was shocked to discover when the names were revealed that my exam had been written by a woman, because it had been so rigorously argued. In another conversation, he suggested that I was doing my required translation in Russian only to please my boyfriend, a professor of Russian. Still, I swallowed all of this. I had learned to expect and endure such sexist comments, and I had been taught, along with many women of my generation, that being nice to people, trying not to expose their failings, particularly if they were men, was more important than standing up for myself.

Then one day this professor jovially instructed me that it was "time for class, dear" and patted me on my rear end at the open doorway of a classroom full of other students, mostly male. My impulse, after I had run down the hall in humiliation, was to tell him how degrading that gesture had been to me, with what economy and precision he had reduced me, in front of my colleagues, from fellow philosopher-in-training to . . . to what? I'm not sure I can say exactly what, perhaps to a child, perhaps to a piece of meat, perhaps to a being so inconsequential that her personal boundaries and integrity were irrelevant. But when I tried to tell him how I felt, the professor laughed and told me that I ought not to be so sensitive, a further humiliation that I didn't even try to explain to him.

Instead, I sought understanding from my fellow students. My closest friends were men; they told me, "Well, what did you expect? You don't exactly dress like a nun!" Even in those more naive times, I knew that my friends were wrong, that my professor's gesture, although it involved physical contact with my bottom, was not an unwelcome sexual advance. That part had taken place weeks before, when I had said "no" and thought little of it. What was going on now was my professor's attempt—conscious or otherwise—to put me back in my place. My transgression had been that I had said "no" to him, in a self-possessed and confident manner. I was a student. I had been counted on to defer—if not to accept, at least to be rattled or flattered—and instead I dared to assert myself as a grown-up woman, as an equal. His response was remarkably effective in restoring a balance of power more comfortable to him. With one eloquent butt slap—performed in front of a roomful of men, that was essential—my arrogant assumption that I was his equal was stripped away with a flourish.

The same dynamics, if we believe Anita Hill (as I and as most people do by now), were at work when Clarence Thomas shoved pornography under her nose. When Hill made her discomfort at his talk of oral sex and penis size clear to him, she sensed "that it urged him on, as though my reaction of feeling ill at ease and vulnerable was what he wanted." Thomas's coin was sex talk. But what made his behavior harassment was his refusal to stop talking dirty, even after Hill, a subordinate, had made her discomfort clear. Hill's account (and Angela Wright's, whose testimony never made the light of day) emphasized that although Thomas may have begun with hopes of romance, the worst harassment occurred—as it did with my professor—after these hopes were dashed. The real spice of the harassment, both Hill and Wright believe, was not a fantasy of overcoming their reluctance, but the pleasures of unnerving and humiliating them. As Angela Wright's friend Rose Jourdain saw it, Thomas's aim was not sex but "making people uncomfortable, you might say harassing them. I think the only thing he was in love with was the conquest of power."

It's not hard to imagine how difficult it would have been for Thomas to tolerate Anita Hill's reserve and self-composure, which were evident throughout the hearings and which I'm sure were evident around the office as well—until she was rattled by Thomas's harassment. One of the classic cultural privileges of manhood, withheld historically from black men, is to be looked up to by the women in one's life, to dazzle them with your masculine authority. When Hill refused to be dazzled by Thomas, the affront was probably even greater to Thomas's manhood than my refusal to be dazzled was to my teacher's. Certainly, throughout the hearings, Thomas's need to maintain a proud and controlled presence (a "cool pose," as sociologist Richard Majors would call it) seemed to go far deeper than the exigencies of politics. Manhood was at issue, as black female supporters of Thomas recognized, some criticizing Hill for having exposed and humiliated him in front of the nation.

Movies like Barry Levinson's *Disclosure*, a provocative role-reversal drama which draws on the sex-fiend conception of the harasser ("no, no, no . . ." Michael Douglas whimpers, forced to have oral sex performed on him by Demi Moore) and makes her a power-hungry executive to boot, have tried to make gender an utter irrelevancy to conversations about harassment. But while women can and do harass men, there are still reasons why it's usually the other way around, and reasons, too, why many women simply accept it. With male privilege—which not all men possess equally, of course—comes a sense of implicit (and often unconscious) ownership of public space and its definitions and values—a sense of ownership that women typically do not feel. When that implicit ownership is challenged by women, when women claim the right to share the power to define and control the rules of the game, sexual and otherwise, men may feel baffled and uncertain about the new rules. They may also feel threatened by the loss of manhood implied in relinquishing the right of sole ownership of public space to women.*

Many women, on their part, put up with and attempt to ignore demeaning behavior not only because they need their jobs but also—as was true in my own case—because femininity demands protecting male egos and subordinating the needs of self to the care of others. Such notions about femininity, although challenged by Nike ads and calls to "power feminism," are still very much alive in this culture. (Although the protection of black manhood is undoubtedly a special case, it's worth considering Thomas's female defenders as responding to their gender training as well as to the need for racial solidarity. Hill's dignity, we should remember, was at stake too, and Hill is no less black than Thomas. Why should Hill but not Thomas have been required to subordinate her interests to a larger community good? Would the same have been required of another black man bringing forth evidence that Thomas had been racially or physically abusive to him?)

For all their racial differences, Thomas's actions and those of my professor had something essential in common, both with each other and with those welders discussed earlier. Their behavior—even when it involved sexual gestures and remarks—was not the behavior of men confused about the rules of sexual courting, or the excesses of sex addicts, unable to control their hormones. Their actions were those of gender bullies, trying to bring uppity women down to size, trying to restore a balance of power in which *they* were on top.

---

*It's interesting how different these cultural changes appear to different parties. In one experiment, psychologists measured the actual amount of time a male and female expert spoke during a national television show; they found that the male took up quite a bit more airtime than the woman. Then, they asked the audience for *their* estimation: Did the man and woman speak an equal number of minutes? Did one speak more than the other? Virtually all the respondents felt the woman had dominated the conversation.

how we got to where we are

How did we come to equate sexual harassment with grabs and gropes and poor libido control?

The sexual emphasis, to begin with, is embedded in sexual harassment law. The Equal Employment Opportunity Commission's 1980 guidelines define harassment as "unwelcome *sexual* advances, requests for *sexual* favors and other verbal or physical conduct of a *sexual* nature" (emphasis mine) when they are either conditions of employment or so much a part of the work environment that they interfere with a person's ability to perform the job. In 1986, the Supreme Court ruled that sexual harassment, when it is "sufficiently severe or pervasive" to create "a hostile or abusive work environment" is a form of sex discrimination, but did not complicate or qualify the EEOC's exclusive emphasis on sexual behavior.

Many commentators have blamed the feminist police for inscribing hysteria about sex into our attitudes and policies regarding harassment. "Feminist excesses," Camille Paglia wrote just before Wright's ruling, "have paralyzed and neutered white, upper-middle-class young men." Well, we apparently didn't succeed with Clinton, who's even married to one of us. It is true, though, that there is a strain of evangelical zeal in American feminism, as there is in virtually every American reform movement. Most of those zealots, however, have rarely been intrinsically anti-sex so much as attuned to power abuses and imbalances that *use* sex as their medium. At the heart of "personal politics" is the recognition that the most intimate arenas of everyday life—even usually pleasurable ones like sex—can function to maintain inequality. If we have sometimes gone overboard, it's been in carrying that excellent insight a bit too far.

Catharine MacKinnon, for example, argues that so long as "what men learn makes them 'a man' is sexual conquest of women" and "women's femininity is defined in terms of acquiescence to male sexual advances," then heterosexual relations are a primary mechanism through which male dominance and female subordination are maintained. If so, then a man's sexual overtures or gestures in the workplace or educational environment automatically carry at least a whiff of intimidation with them. Some feminist agendas—hardly the majority, contrary to what Camille Paglia and others have suggested—have been inspired by this sexually essentializing view of men, and also of harassment itself. Ellen Willis writes:

"Some recent harassment complaints sound like heavy-handed satire: a male professor is charged with using a sexual comparison to make a point in class; another is enjoined from keeping a picture of his bikini-clad wife on his desk; a female professor demands the removal of Goya's "Naked Maja" from her classroom wall; a teaching assistant, supported by her professor, warns a student she considers him a harasser for handing in a paper containing an 'inappropriate' sexual analogy . . . Can covering piano legs be far behind?"

The problem with thinking of things in MacKinnon's terms is that it converts a (pretty accurate) description of a still-dominant sexual *ideology* into a much too general description of the contours of *actual* relations between men and women. Much sexual ideology indeed naturalizes male aggression. "Birds do it, senators do it, even fuzzy little bees do it: they engage, that is, in the ancient art of sexual harassment . . . [it's] probably as old as the partition of sex cells into sperm and egg"—that from the science section of *The New York Times*. Numerous Hollywood comedies "define" men this way too by rewarding the intrusive, persistent suitor—the man who won't take "no" for an answer—with the eventual affections of the heroine. (Some, like *Something About Mary* and *You've Got Mail*, even appear to justify male stalking.)

But whatever pop science, pop psychology, and the movies are serving up to us about "Real Men," not all *real* men are buying it, or acting in accordance with it. Not only are many men temperamentally unsuited or morally disinclined to play the role of sexual aggressor or conqueror, but our culture is sending out some mixed messages about those who do. Being a male animal, as we've seen, is just one wrong move away from being seen as a disgusting predator. In such a cultural climate, many men—far from feeling permission to invasively grab and grope—have felt that the safest course is to avoid all sexual references and compliments, all affectionate touching and glances. This is a tall and grim order, especially for those who have grown up in families or ethnic communities that use touch to communicate empathy, humor, and playful mutual recognition.

The Thomas/Hill hearings, for all that they raised consciousness about sexual harassment, also contributed to a distorted view of its nature. Here, the blame is not to be laid at the door of feminists, but with Republican defenders of Thomas, who cunningly used moral outrage over pornography to raise disbelief that Clarence Thomas could possibly be guilty of such disgusting crimes. Recall Orrin Hatch's pious displays of shock over Long Dong Silver and pubic hairs on Coke cans, his overheated speeches about how only a "psychopathic sex fiend" would show pornography to a subordinate. The strategy served Hatch's political purposes admirably, since it made it a requirement of finding Thomas guilty that one should acknowledge him a psychopath, which I don't think even Anita Hill believed he was. In the process, a narrative about an alleged abuse of power was replaced by a lurid story of alleged *sexual* obsession. Exploiting the confusion thus created between the thrill of intimidation and the thrill of sex, it was frequently argued by defenders of Thomas that he couldn't possibly have been sexually interested in Hill because of the other romances he was involved in at the time he was supposed to have harassed her. The implication here is that if Thomas was "getting enough" to satisfy his sexual needs elsewhere, he wouldn't have needed to harass Hill.

Arguably, the senators' outrage wasn't just political strategy; it also enabled them to create a singularly obsessed, grotesque devil—the sex fiend—onto which all the crimes of male sexual excess could be projected. The same thing has happened throughout the Clinton scandals. "Despicable behavior!" (Larry King). "It's disgusting . . . not the way normal people act" (Representative Thomas Davis of Virginia). "So shocking and so sordid . . . Humiliating . . . And with an intern half his age . . . A tragic and pathetic tale!" (Lou Dobbs, on *Moneyline*). Such proclamations of revulsion, often, have tumbled out of the mouths of men whose own sexual record is far from immaculate, some—like Henry Hyde—who engaged in far more serious extramarital affairs than Clinton's. (I bet the sexual practices of some faithful husbands too, if subjected to the kind of investigation Bill's was, would yield some "abnormal" little practices.) Calling the kettle black, it seems, allows the pot to imagine (or pretend) it's stainless. *He's* a "shameless and arrogant liar and predator of women" (as one commentator described Clinton); the implicit message: my outrage proves *I've* never cheated or lied myself.

The media also deserves a large share—perhaps the largest share—of blame for the sex-fiend portrait of the harasser. The complexities of

institutionalized inequality don't make for riveting television or reporting. Dirty details do, whether or not they constitute a crime.
Remember the Packwood affair, and all those diary excerpts, among
which the press made no attempt to sort out the salacious, sopho-
moric, and merely silly ramblings of a man who gushed like a teenager
over his (consensual) sexual belt notches from those behaviors of his
that were invasive and bullying. *"Diary reveals an obsession with sex
life,"* reads an early headline. When did that become a crime?
Newsweek published extensive excerpts from the diaries, which included passages such as the following (I've included the *Newsweek* reporter's editorial comments):

> On Nov. 21, 1989, Packwood began his diary describing a staffer he re-
> ferred to as "S-1," a "very sexy thing" with "bright eyes and hair and that
> ability to shift her hips." After a few glasses of wine that evening, "I finally
> said to her, 'S-1, would you like to dance?' She says, 'I'd love to.' So I
> slipped around the side of this gigantic desk and we danced. . . . Well, I
> won't bore you with all the details of the evening. S-1 and I made love, and
> she has the most stunning figure—big breasts . . . They stand at attention."
> As Packwood was lying with S-1 on the floor of his office, she said, "You
> have no idea the hold you have over other people." "What is it?" asked
> Packwood. "Well, I think it's your hair," she replied. "We both laughed,"
> recorded Packwood. "Now bear in mind that this is an hour and a half after
> we've made love and we're both still nude and lying on the rug." Packwood
> gleefully observed ("Get this," he wrote) that two of his aides were right
> outside the door the entire time. Packwood seemed quite pleased with his
> hairdo. On March 20, 1992, after 20 minutes in the hot tub, he recorded, "I
> just blew my hair. I didn't use any gel on it at all. I just blew it until it was
> about dry, combed it, and if it didn't come out looking just right. It had just
> the right amount of bounce to it, and wave to it."

There was plenty of real harassment in the Packwood case, which
those titillating but irrelevant passages in which Packwood recorded
his admiration of perky breasts and sexy hips confounded. Packwood,
we know, did a lot of aggressive invading of the bodies of staff assis-
tants, hostesses, and elevator operators. But having sex with a woman
who felt relaxed enough to tease him about his hair does not seem to
fall into quite the same category. Surely writing gushingly in one's own
diary about bodies to which one is attracted, no matter how crude the
aesthetic, is not sexual harassment. And what abuse is involved in

Packwood's pleasure at having gotten his hair to have "just the right amount of bounce"—"after 20 minutes in the hot tub," as we are reminded? Why include such passages, except out of prurient fascination with the sexual details or giggling delight at exposing Packwood's lack of masculine stature? The (unacknowledged) "crime" Packwood commits here is to expose a narcissism that our culture has generally deemed unmanly, feminine. These passages are embarrassing precisely because Packwood seems in them not like an adult man, but a perpetually excited adolescent boy who can't get over the fact that he is actually having sex with women.

Many of the same dynamics were at work in reporting on the Jones affair and then—out of control, unimpeded—the Clinton/Lewinsky affair. Starr couldn't have gotten anywhere without the media's enabling. American sanctimony is matched only by the pleasure and profitability of ferreting out the same sexual secrets that we drive underground. It's in the context of this double-sided coin of salacious Puritanism that the Republican moralists and much of the mass media have been in perfect synergy over the Clinton affair. Many of the worst media offenders, remarkably, continue to see themselves as "liberal." They are too naive to realize (or too self-serving to care) that a witch-hunt can occur without its being fully orchestrated—or even always intended—by all the participants.

is bill clinton a gender bully?

It was thoroughly predictable, once Starr's avalanche of illicit but consensual acts had buried all possible hope of the nation's distinguishing between sneaky sex and criminal sex, that Paula Jones would claim "vindication" by Starr's report, as she began her appeal* in October of 1998. It was the most unjust of ironies. The Starr investigation into the President's sex life would never have *happened* but for Jones's sexual harassment suit—which was dismissed in April. Six months later, without blinking, Jones's lawyers invoked Starr's report as evidence of the merit of their case, in the hope that the court of appeal would reinstate it. See, we were right, they gloated—he *is* a dirty scumbag. But

---

*Ultimately, the case was settled out of court.

whether or not Bill Clinton is a dirty scumbag, of course, hardly answers the question of whether or not he's ever been a sexual harasser.

There is certainly nothing in Starr's report—or, it appears, in the deluge of documents that followed it—to suggest that Clinton sexually harassed Monica Lewinsky in the "grab and grope" mode. By all accounts, she came at him panting. That in itself doesn't settle the issue completely for me, however. "Welcome" and "unwelcome" are not the most morally or pragmatically useful categories to employ when a subordinate's sexual feelings for an authority figure are at stake. Even as a middle-aged college teacher with hardly the dazzle quotient of a President, I know that a student's "crush" can blossom into a dangerously lopsided state of affairs—as it appears to have with Clinton and Lewinsky. At the very least, Clinton seems to have been extraordinarily naïve about the kind of *emotional*—not to mention political—trouble that can walk in the door when a boss and a subordinate, or a teacher and a student, have an affair.

Being naive or careless is a far cry, however, from deliberately exploiting someone. William Bennett, pretending to be a feminist, said that Clinton treated Monica as a sex toy, and never really thought of her as a human being. Bennett counted on that image of Clinton on the phone, being "serviced" by Monica while he does business, resonating with his—Bennett's—description of the relationship. But who is leaving Monica's desires out of the picture here, Clinton or Bennett? From all accounts, Monica *loved* oral sex (giving the lie to all those jokes about Jewish princesses), *begged* Clinton to allow her to perform it to its conclusion (which he resisted), and may well have derived a thrill from exercising her own particular brand of "power" on Clinton while he conducted high affairs of state.

Indeed, a thoughtful reading of the current information on the relationship suggests that Clinton, far from being the intimidating, harassing variety of male animal with Monica, played rather the (traditional) role of "woman" in the relationship. He seems, first of all, to have been *seduced*—by Lewinsky's flattering attention and sexual potency (yes, women can be sexually potent too). Second, despite his giving in to her advances—and ultimately becoming more active in the relationship—he remained an ambivalent participant throughout, holding out on expressions of intimacy that had deep meaning for him, and frequently expressing his qualms about what was happening. Third, a good deal of his behavior seems to have been about "being nice" to Monica, making her feel she was important to him, not hurting her feelings. (He finally allowed her to bring oral sex to conclusion, she reports, because he "didn't want to disappoint her"!)

From the beginning of his presidency, Clinton has been dogged by a reputation for being too soft, sentimental, and undisciplined. Certainly, he didn't behave like a lean, mean machine with Monica. After he tried to end the affair yet she continued to pursue him, often getting quite demanding and intrusive, he wasn't nasty or imperious with her, but tried to let her down nicely—with presents, physical tenderness, humoring her with vague allusions and weak jokes about a possible future together. To my mind, *those* were the biggest lies Clinton told. Do I know for sure that this was how the situation played itself out? Of course not. Nonetheless, I have a certain confidence in my interpretation. Why? Because these are the same kinds of lies that I—and I imagine I'm not alone—have told people with whom I've gotten more entangled than I wished to be, but whose feelings I cared about, and who I *hoped* would not really believe me, but feel appreciated enough to go away without gaping wounds.

Bill Clinton, then—unlike my teacher or Clarence Thomas—hardly seems to have been a gender bully with Lewinsky. Yes, there were troubling aspects to the White House's treatment of the intern. From my perspective, the most chilling line in the Starr report referred to an action—not taken by Clinton but by Evelyn Lieberman—that has been widely praised by the press. "I decided to get rid of her," Lieberman said about booting Monica Lewinsky out of her White House job. She did it to protect Clinton, of course, so in that sense you can see it as Clinton's "fault." But Clinton himself, having found out about Lewinsky's dismissal from the White House, tried to repair the damage. That's the context in which I interpret the ensuing commerce in jobs and so forth—not as an attempt to obstruct justice by buying Lewinsky's silence with a juicy new position, but as an attempt to right the wrong that was done to her when she was shuffled around purely to protect the President. By her own admission, she viewed Clinton's willingness to help find her a job on her wish list as a way for him to demonstrate that he felt bad and wanted to compensate her for the dislocation and hurt she suffered.

If William Bennett had really been interested in the treatment of women at the White House, he would have concerned himself with other things besides Clinton's sexual behavior. He would've wanted to know: Are women employees there denigrated, taunted, made to feel inferior or incompetent? Are they relegated to lowly or sex-stereotyped activities, patronized, rewarded for being cute and having big breasts and ignored if they are less attractive? Are the female interns excluded from social networks and invitations extended to the males? William Bennett never showed any interest in questions such as these, and in the wake of the Lewinsky affair neither has anyone else. But the answers to them are far more important to feminists—male and female—than whether Bill Clinton committed a high crime when he defined sexual relations to exclude the blow job.*

The exigencies of partisan politics can sure bring strange new recruits to the feminist fold. Apparently, as Clinton's troubles moved into the impeachment phase, a number of his Republican enemies suddenly realized, after careers of belittling women's issues, that sexual harassment was a serious crime indeed. Paula Jones's "civil rights" were being trampled! Clinton had attempted to "obstruct" the "justice" due her! Never mind that Jones's case had been dismissed. Never mind that Jones's case would not have been advanced by Clinton's acknowledging a consensual affair that was admittedly instigated by Lewinsky herself. Never mind that just eight short years ago three gallant new defenders of the rights of Paula Jones—Arlen Spector, Orin Hatch, and Henry Hyde—had been major players in the attack on Anita Hill.

---

* Please note: I would never call for the Starr Chamber or any office like it to turn its unchecked power to examine gender discrimnation in the White House. Indeed, I believe (as many others do) that the office of the independent counsel is itself an abuse of power, and should be abolished.

beautiful girls, from both sides now

Scene I: A woman sits in the car next to mine, stopped at the light. She flips down her visor, on which is mounted (as it is in my own car) a small mirror. With all the deftness of one who has been doing this all her life, she quickly applies fresh lipstick, then—out of the corner of her eye—sees that I have been looking at her. Brief embarrassment, and then mutual recognition. We exchange a smile in which is contained a voluminous and intimate knowledge of each other. She is attractive, thirtyish, athletic, looks a bit like Christie Brinkley. But in her face I see—and she knows I see—the little girl who once played at dress-up, lipstick smeared, eyes flirting at herself in the mirror, supplying the absent male with her own gaze, practicing. I see—and she knows I see—the fifty-year-old woman to come, scrutinizing every line, carefully applying her lip pencil just outside the line of her lips for fullness, shadow to lift the corners of the eyes, wondering what she really looks like now that men are no longer announcing it to her with a comment, a look, a smile. We both know that whatever and however much else we accomplish, we will continually ask ourselves the same beseeching question. We know that no matter how young or slim or beautiful we are, we will need to ask it again and again, ask it of the eyes that survey and judge us, begging for a reassurance that can never be secured.

Different words, the same question, year after year. "Am I pretty? Am I too fat? Am I getting old?"

Scene II: A woman sits in the car next to his, stopped at the light. She flips down her visor, on which is mounted a small mirror, and applies fresh lipstick. She is thirtyish, attractive, blonde. She looks really put together. There's something a bit like Christie Brinkley about her— or maybe he's just mixing up real life with that scene in the movie with Chevy Chase. In any case, she's blonde, and pretty, and looks really athletic, and she's wearing sweats like she's just come from the gym. Just looking at her makes him kind of nervous, makes him feel young and awkward. This is a girl straight out of the Nike ads, the kind of girl he and his friends stare dumbly at when she walks past them in the bar, then roll their eyes to each other. Cool, self-assured. The kind of girl he'd never, ever have the nerve to ask for a phone number. The deft way she's putting her lipstick on (how do women do that so quickly, so surely?), you know she's got a lot of self-confidence, knows how to take care of herself. She catches his glance from out of the corner of her eye . . . and smiles! Sort of shyly, embarrassed . . . or maybe it's flirtatiously, he can't exactly tell. Did he see correctly? He steals another glance at her to check. She blushes, and smiles again. Is she sending him a signal? Is something expected of him? Should he do or say something?

The light changes and she speeds away, leaving him vaguely humiliated, unsettled, messed with.

our bodies, ourselves in an empire of images

I've always found it difficult to convey to men, without seeming to hyperdramatize or lack a sense of proportion, what it's like to be a woman in this culture in which images of flawless female bodies are everywhere. One reason for my failure, I've come to realize, is that the tales of women's insecurities, obsessions, disorders are only half the story. It's been mind-opening for me as a women to think about the question of women's beauty from the "other side," in terms of its impact on men. Studying that impact, I've come to see that it's dead wrong to think of men and women as creatures from different planets, hard-wired with incommensurable responses, values, styles. Yes, we're different in many ways, and some of those differences may be biologically grounded. But many, too, are—paradoxically—the result of what we share. We are immersed in the same culture, haunted and taunted by the same images. They simply do their work on us in mirror-image ways.

Take, for example, our ideas about what constitutes an acceptable female body. It's only common sense to expect that the bodies of movie stars and models would begin to function as norms of male perception as well as female. These images alter our very notions not merely of what's "beautiful" but also of what's "normal" in a female body. Mass-media images, for example, have altered notions of what breasts are supposed to look like: glorious globes standing at attention even when supine are what boys now grow up expecting to see when they unhook their first bra. In *L.A. Story*, Steve Martin asks Sarah Jessica Parker why her breasts feel funny to his touch. She replies, "Oh, that's because they're real." It's hilarious but true that real breasts are the anomaly in visual culture today; it's rather a shock when a naked actress lies down and her breasts flop off to the side. It doesn't look *right* anymore.

Where visual culture goes, so ordinary folks eventually follow. I wouldn't be surprised if implants don't soon become the de rigueur Sweet Sixteen gift among prosperous middle-class families (as nose jobs once were in certain ethnic neighborhoods). Breast augmentations are already the most popular form of cosmetic surgery; we're becoming a culture of Stepford Breasts! It's not because women are vain creatures who won't be satisfied unless they look like goddesses. Most of them merely want to look *normal* in a culture where "normal" is being radically redefined by the surgeons, the actresses, and—of course—the aesthetics of the media. Redefined not only for them but for their boyfriends and husbands too. "Most movies today," screenwriter Richard Donner says, "are green-lit off the hormones of a twelve-year-old boy." Donner has tried, with minimal success, to convince directors to bring female characters to the screen that are not (as he puts it) "collagen-lipped, balloon-breasted grotesqueries of femininity." But while grotesque to Donner (bless his heart), those collagen-lipped, balloon-breasted images are fast moving beyond the realm of hormone-driven fantasy, becoming instead those twelve-year-old boys' blueprints of the female body.

This is an area that has not yet begun to be adequately researched. Think, for example, about the issue of men's attitudes toward fat. While psychologists have paid some attention to the impact of unrealistic images of the female body on women and girls, virtually no attention has been paid to their impact on men. In that area, our knowledge is composed mostly of folk wisdom—increasingly anachronistic folk wisdom. A widespread cultural rumor exists that men are more tolerant about women's fat than women think they are, and that the woman who obsesses about her weight is thus the victim of a paranoia induced by gay fashion designers and other women. On more than one occasion, I've had men in the audiences at my talks protest that as far as body-image issues are concerned, women are their own worst enemies. The belief that *men* want thin women is all in our heads. Black men have sometimes charged me with a racialized view of beauty, one which overlooks the flesh-celebrating aesthetic values of nonwhite cultures. Skinny women needn't apply *here*, in our neighborhood!

It's true that there are ethnic traditions—my own Jewish heritage included—that have held the zaftig (juicy, voluptuous) female body in greater regard. But each of us also lives within a dominant culture that often recoils from these "alternative" ideals. Black and Jewish women may grow up sharply aware of representing for that dominant culture a certain excess—of body, fervor, intensity—which needs to be restrained and, in a word, made more "white." The men who share our ethnic backgrounds are not immune to these associations, or to the media images which equate female beauty with slenderness. Truth be told, I find African-American and Jewish men today to be even slightly *more* fat-phobic than the average white guy. They may revere their plump ethnic mothers, but few men want such an unassimilated body on their arm when they go to a business dinner or attend a party at their boss's house. The men who protest at my talks may find it uncomfortable to face, but things have changed. I teach a younger generation of men, and I *know* things have changed. With few exceptions and whatever their race, they want girlfriends who are trim, tight, and toned. Moreover, older men seem to be following suit. Even the want ads in the intellectual *New York Review of Books* specify "trim" and "slim" among their requirements!

Princess Diana's bulimia began, by her own account, the week after she and Charles got engaged, when Charles put his hands on her waistline and said, "Oh, a bit chubby here, aren't we?" I've looked hard at many photographs of Di from the time of this remark, and I see a round face but nothing remotely like what I would call a "chubby" body. But "chubby," of course, is a matter not of weight charts but of perception. And men's perceptions, no less than women's, have definitely been vulnerable to the shrinking cultural standards. It's true that most men *don't* go for skeletal, gaunt bodies, a fact that they make much of. When I put slides of Kate Moss up during my talks, the men groan ostentatiously, proud to be able to say that she turns them off. I find their desire to reassure the women touching, but flimsy—and the women know it. They know that just because men don't appreciate Auschwitz chic doesn't mean they are "tolerant" of fat.

It's easy for women to get furious at Charles for his remark to Di, as I got furious many years ago with the man in my class who wrote in his journal that his (quite slender) girlfriend's "thunder thighs" were revolting to him. How, he asked me, could he get her to go on a diet? My anger was short-circuited, however, when I realized that just beneath the surface of his revulsion was tremendous anxiety. Anxiety that with a less than perfect girl on his arm, he would look like a loser to the other guys. Anxiety that if she got fat, he would stop being turned on by her. I was deeply pained for his girlfriend, with whom I strongly identified, and who had written in *her* journal about being goaded by him to lose weight because she weighed more than Brooke Shields. *His* anxiety, however, reminded me that the perfect images of beautiful women are setting men up for obsession and failure too. For just as the beautiful bodies subject us women to (generally) unrealizable models of the kind of female we must *become* in order to be worthy of attention and love, they also subject men and boys to (generally) unrealizable models of the kind of female they must *win*—with equally destructive consequences.

All his life, a man is bombarded by these images, "intruding" on his consciousness (as author and activist Timothy Beneke aptly puts it), and "designed to distract, arouse, and awaken male sexual feeling." A boy can avoid (if he is willing to be a cultural hermit) buying the magazines, turning on the television, going to the movies. Even so, there are the billboards, the ads in the subway, the shopwindows. But boys, unlike girls, are often not very cognizant of the power the images of female beauty have in their lives. When I first talk about these issues in my classes, there often seems to be a gaping disparity—or at least so it appears on the face of it—in male and female beliefs about *men's* attitudes toward images of female beauty and sexuality. The women in my classes claim that the men are profoundly affected by them, and feel themselves evaluated and judged on the basis of their success or failure in conforming to them. The men protest that the images belong to the realm of fantasy, with little bearing on their expectations of "real" women.

A girlfriend and boyfriend who were taking my class together almost split up over the Farrah Fawcett poster that hung over his bed in his dorm room. She couldn't get undressed, she told the class, without feeling that Fawcett's perfect breasts were mocking her own. *He* insisted that Farrah belonged to "a different world" than the world he inhabited with her. These discussions got nowhere so long as our focus remained on convincing him of the legitimacy of *her* feelings. But when we turned to explore his, doors began to open. He began to see that fantasy and reality—in least in our culture—cannot be put in separate, self-contained little boxes.

A pair of twin messages work together in perfect, if destructive, synergy in the lives of males and females. Here, the images say to men, is what you can have—if you're forceful, successful, manly enough. The Prize for being a Man. Desire her. Dream about her. Go out and get her. You're nothing if you don't. At the same time, and at increasingly early ages, the girls are trying to *look* like the girls in the Nike ads and teen soap operas, trying to become the fantasy, the object of male desire. The trouble they run into—by way of low self-esteem, eating disorders, exercise compulsions, body-image problems—is well documented. This is not a success story for most girls.

But let's follow the narrative from the boys' point of view too. In *their* version, some girls *do* succeed in becoming the ideal—or pretty close to it, close enough for a teenage boy anyway. Such a girl becomes—in the world of adolescent men—what Warren Farrell calls a "genetic celebrity." Simply because of what she looks like, she achieves a glittering status, she's the incarnation of the Prize. Date her and you strut down the hall, proud and tall. But this is not a success story for most boys either, for these girls are out of reach for the ordinary boy:

Next to the genetic celebrity . . . he, a fourteen-year-old, pimple-faced, bumbling adolescent, feels as nervous as a groupie. [And he] notices something about the genetic celebrities in his ninth-grade class: They are going out with the eleventh-grade boys. He does not feel equal to the most attractive girls in his class. He does not feel equal to his peers . . . The genetic celebrities might be willing to go out with him if he earns his way to their attention by performing as a football player or class president. He feels desperate.

But why is he so desperate? Why doesn't he just settle for a friendly, pleasant-looking, noncelebrity girl? Ultimately, of course, most heterosexual boys do, but many remain haunted by the beauties. Here we come to the deeper, more disturbing aspect of it all. It's not just a matter of impressing the other boys or proving that one is a man. The images—of breasts and legs, of eyes and mouths, of Farrah and Cindy and Tyra and others—not only shape perception, they also shape sexual desire. Straight male sexuality is honed on the images, even fixated on them. "Anything less," Farrell says, "feels like an inferior fix."

As someone who was one of those "inferior fixes," I looked on the genetic celebrities of my school classes with a deep, mean, sickening envy. The first was Karen Gross, whom I apparently tried to write a story about. I found the following in a stack of old papers:

What a perfect package she was—as well decked out for Troy Donahue vintage adolescence as I felt I was an anomaly. She was a Jewish Annette Funicello, skinny enough to buy her clothes at S. Klein in downtown Newark. A desirable "good girl," the boys swarmed around her for the little that she made available. The mystery kept me up at night. What was it? Which combination kept them magnetized? I scanned *Teen* magazine to help me. Shining eyes? Clean hair? Fresh breath? I pictured her on couches, rejecting advances and being loved for it. Donna Kantrowitz, who wore ballet slippers to school and after class, it was rumored, laid down on a park bench and got it repeatedly from the football team, had dates, of course. But Karen was different. She was *popular* in the Friday night sense. Just a kiss from *her* was enough.

Later on, in high school, the genetic celebrities were the ones who looked like Judy Collins and Mary Travers, not Annette Funicello; I didn't even have the right kind of hair, let alone the right kind of body. I tried desperately to offer other things—my satirical wit, my rebelliousness, my intelligence—to the guitar-playing poets and politicos whom I hung out with. But although some did ask me out, it was very clear their fantasies lay elsewhere, and there was simply nothing I could do about it. I wasn't the right type.

Looking back to that time, I remember being in continual low-level pain over that recognition of not being "the right type." As Farrell points out, these are primary—and perhaps primal—fantasies and rejections we are talking about here. They are deep, and they are hard to dislodge. We can spend a lifetime getting even for past hurts, trying to rewrite our adolescent histories. Farrell provides this painful account of Neil Diamond in live concert:

> Encore! Encore! Sixty thousand people stomping their feet, flooding the aisles. Entranced. Enamored. The man focuses the microphone before his powerful lungs. But there is no song. "Here I am, Kathy. The pimple-faced boy who sat behind you in seventh grade . . . Here I am. Neil—who wasn't good enough to kiss you after class . . . the class frog . . . Well, your frog has turned into a prince, Kathy. Kathy, wherever you are—they're eating out of my hands, Kathy . . . eat your heart out now, Kathy . . . eat your heart out . . . wherever you are . . ."

When the boys in my class moaned and groaned about *their* unrequited crushes on the genetic celebrities in our class, I had some sympathy. But I was mostly immersed in my own suffering, which I considered intractable and out of my control, while theirs seemed superficial and their own bloody fault. Today, I realize that their obsessions were as little in their control as mine. It seems such a commonsense insight, yet it eluded me for much of my life. The images, I knew, had *me* hypnotized. I pored over the magazines and record sleeves, trying to figure out how to remake myself. I ate nothing but cottage cheese, put strange objects in my hair, bought ridiculous clothing. But it took me years to "get" the fact that the same images were a part of the boys' world too; I was too competitively fixated on the girls in my classes who had the right look. When I did realize, the object of my rage almost immediately shifted away from other women toward the culture that had us all behaving like sexual robots.

Today, I am especially admiring and appreciative of those male writers who have been honest and brave enough to admit the power of the images in their lives, even—or perhaps especially—when it reveals ugliness and violence in their sexual psyches. Usually, the real ugliness and violence are in the cultural landscape. In *Soul on Ice*, for example, Eldridge Cleaver recounts how when he was in prison he decided to paste a pinup girl on the wall of his cell:

> I would fall in love with her and lavish my affections upon her. She, a symbolic representative of the forbidden tribe of women, would sustain me until I was free. Out of the center of *Esquire*, I married a voluptuous bride. Our marriage went along swell for a time: no quarrels, no complaints. And then, one evening when I came in from school, I was shocked and enraged to find that the guard had entered my cell, ripped my sugar from the wall, torn her into little pieces, and left the pieces floating in the commode . . . Almost every cell, excepting those of the homosexuals, had a pinup girl on the wall and the guards didn't bother them. Why, I asked the guard the next day, had he singled me out for special treatment? . . . "Tell you what," he said, smiling at me . . . "I'll compromise with you: get yourself a colored girl for a pinup—no white women—and I'll let it stay up. Is that a deal?"

The incident could have been taken by Cleaver simply as evidence of the racism of the guard. Instead, Cleaver went through a period of painful self-questioning, in which he realized that he had *chosen* the white girl over available pictures of black women, that he really did prefer white girls to black. "It's a sickness," one of his jailmates tells him. "All our lives we've had the white woman dangled before our eyes like a carrot on a stick before a donkey: look but don't touch." Of course, Cleaver's jailmate was right. Black men have been lynched for merely catching a glance at white women. In 1955, while Cleaver was in jail, a fourteen-year-old Chicago boy named Emmett Till, visiting relatives in Mississippi, allegedly whistled at a young white woman working at a drugstore where he had gone to buy chewing gum. That night, the woman's husband and another man came to where Till was staying and took him away for "questioning." Two days later, Till's battered body, shot in the head, was found in the Tallahatchie River. Till's accused murderers were acquitted by an all-white jury.

Unlike his jailmate, Cleaver finds his dawning recognition that he has been "indoctrinated with the white race's standard of beauty" to be too disturbing to simply accept. One day, looking at a photo of the white woman who had caused Emmett Till's death, he realizes that he feels desire even for *her*, and he snaps:

> I looked at the picture again and again, and in spite of everything and against my will and the hate I felt for the woman and all that she represented, she appealed to me. I flew into a rage at myself, at America, at white women, at the history that had placed those tensions of lust and desire in my chest.

Cleaver describes how he advanced to a career as a rapist, first "practicing" on black women, then venturing into neighborhoods inhabited by "The Ogre"—as he came to call white women. When I taught Cleaver's book in class, many of my female students could not overcome their horror and anger over his career as a rapist to appreciate Cleaver's honesty and insight. Even today, I find *Soul on Ice* one of the most intelligent, illuminating discussions of the intertwinings of racism, sexism, beauty, violence, and sexuality in this culture.

beauty, power, and pornography

You don't require racism to elicit male rage over what is perceived as the power of female beauty to invade male consciousness and arouse desire and then to reject that desire, leaving the man humiliated, shamed, frustrated. Timothy Beneke has collected dozens of chilling interviews, some of them with rapists, in which he explores how "intrusive images" of beautiful women create a cultural climate which arouses male resentment and anger. From one of those interviews:

> To be honest, it just makes me hate beautiful women because they're using their power over me. I realize they're being used themselves, and they're doing it for money. In *Playboy* you see all these beautiful women who look so sexy and they'll be giving you all these looks like they want to have sex so bad . . . In this society, if you ever sit down and realize how manipulated you really are it makes you pissed off—it makes you want to take control. And you've been manipulated by women, and they're a very easy target because they're out, walking along the streets, so you can just grab one and say, "Listen, you're going to do what I want you to do," and it's an act of revenge against the way you've been manipulated.

The rapist's response is an extreme one. But it is not only the rapist, Beneke suggests, who feels "powerless, distracted, and resentful" of beautiful women. From another interview:

> Let's say I see a woman and she looks really pretty and really clean and sexy, and she's giving off very feminine, sexy vibes. I think, "Wow, I would love to make love to her," but I know she's not really interested. It's a tease. A lot of times a woman knows that she's looking really good and she'll use that and flaunt it, and it makes me feel like she's laughing at me and I feel *degraded.*

It's unlikely that the pretty woman is really laughing at this man or wants him to feel degraded. And I don't believe that a woman's "giving off sexy vibes" is equivalent to rescinding all personal boundaries and choices over where particular relationships will lead. But women often *do* dress in order to feel sexually powerful, and we need to acknowledge that fact. Moreover, since women may indeed want to attract men or seek reassurance from them through their sexual power, there is a logic, I'm afraid, to a man's feeling "harassed" by beautiful women at the workplace. Here's Nancy Friday, putting it (as usual) in the strongest, most feminist-blaming terms she can muster:

> Putting sexual beauty into the workplace was dynamite. The Dress for Success Suit was a form of armor that solved many problems, but when sexy fashions and high heels returned to the fashion scene, the people who were left to deal with this unprecedented event were men.
> Women, according to feminist law, were allowed to do whatever they wanted with their bodies and men were forbidden to react. What background did men have in working alongside beautiful women who aroused them, confused them, and many times made them very angry by rejecting them?
> Instead of educating women regarding the reactions we provoke when dressing in an eye-catching way, we give them outrage, the legal weapon of sexual harassment to use against any man who responds in a manner that makes the woman uncomfortable. We might have educated women regarding the language of clothes; instead, feminism declines even to consider that women, consciously or not, may be involved in their own harassment, that there are ways of dressing, walking, talking, that signal to a man something that women may not have intended.

I'm not in agreement with Friday on all points here. I don't share her rather demeaning view of men as sexual time bombs who cannot help but go off at the sight of high heels. And I wonder, too, if she would blame sexy men at the office who arouse *us*, confuse us, and then reject us in the same way as she blames sexy women. Maybe Friday has never experienced such rejection. I have. I would never think of a beautiful man as harassing me simply because he looks, moves, smells sexy, and then has the effrontery to not give himself to me. But then, I didn't grow up with provocative images of semi-dressed men distracting me on every street corner, suggesting to me that they are mine for the taking . . . if I'm woman enough.

The fact is, too, that Friday has a point about women's failure to take responsibility for the sexual signals we may give off with provocative clothing. Don't misunderstand—I'm *not* saying that wearing a short skirt is an invitation to rape. But there is a lot of territory in between blaming a woman for male reactions to her body and us giving ourselves carte blanche to dress as erotically, as revealingly as we like and allow ourselves to think it can "mean" anything we like. Men are dead wrong when they interpret sexy clothing as "asking for it." (No kind of clothing, no matter how revealing, is capable of "asking" for anything; only *people* can *ask*.) But *we're* wrong if we think we can ever have absolute control over just how much of a sexual presence we are projecting. That's where Friday's comment about the desirability of becoming educated about the language of clothes is apt. I'm not talking about regulating what women can wear; I'm talking about women not slipping into bad faith or disingenuous ignorance or innocence about our own sexual power.

Most men, of course, do not turn to rape when they feel degraded or humiliated by the power that they perceive female beauty to have over them. But some may find solace (and excitement) in pornography. Pornography, of course, takes many forms and different people seek different kinds of experiences from it. Some seek in it the thrill of transgression, some are turned on by the impersonality of the sex depicted in it, some are titillated by re-creating the experience of being a dirty, naughty child. Some men may be turned on by images which degrade the *woman* and thus restore men's sense of pride, superiority, and power over the bodies which arouse them. The humiliation of women, however, seems not to be at the core of the pornographic experience for most men—at least, not the soft-core experience.

Many, many men instead describe heterosexual pornography as a fantasyland in which male desire is always welcomed by beautiful women and the male body never rejected, no matter what it looks like or what it does. A "utopian kingdom" (as Philip Lopate calls it), in which the genetic celebrities give themselves freely to even the most ordinary men. As Fred Small writes:

> Economic and political realities notwithstanding, most men do not perceive women as powerless, in part because women hold the power of rejection. I suspect that a man who whistles at women on the street actually perceives women as having more sexual power than he. We are trained from childhood to believe that real men get sex from women, that if we do not get sex from women we are not men, we are nothing. Women can deny sex to men, thereby denying our manhood, our existence. Men do not want to hurt women. Men hurt women only when they have been fooled into believing with all the force of hallucination that they must hurt women in order to save their own lives. It is a brutal order that robs the humanity of both men and women. *Playboy* offers men a dream vacation from this system. In its pages, women are not aloof and rejecting, but welcoming and sexually accessible. *Playboy* delights in showing us "nice girls"—the Girls of the Ivy League, the Girls of the Big Ten—wanting sex, wanting us. In these purposefully airbrushed smiles that speak eager, delighted consent, they are not powerless; they freely surrender their power of rejection. They are not coerced or hurt; they are on our side.

What much heterosexual pornography seems to provide for many men is a world in which women are in a state of continual readiness and desire for sex, but utterly without expectations. The women are voracious, yet at the same time completely satisfied by anything the male has to give and not needful of that which he cannot give. There's no possibility of rejection or indifference. But no possibility, either, that he will fall short, that she will want too much from him. Women in porn are hot, but in a crucial way their subjectivity is *cool*, undemanding, accepting, placid. They provide a fantasy of sexual encounter that does not put manhood at risk in any way.

In the quote above, Small is talking about *Playboy*—very soft-core indeed. But even pornographic motifs that may seem to some women to be quintessential expressions of a male need to degrade and dominate—for example, scenes which depict men ejaculating on women's faces—are, from this point of view, fantasies of unconditional acceptance. As Scott MacDonald puts it:

From a male point of view, the desire is not to see women harmed, but to momentarily identify with men who—despite their personal unattractiveness by conventional cultural definition, despite the unwieldy size of their erections, and despite their aggressiveness with their semen—are adored by the women they encounter sexually.

Pornography thus becomes a context in which the penis, haunted by old guilt and embarrassment about secret masturbation, wet dreams, unwanted erections and ejaculations, taught that what spurts out of the body is disgusting, can come out of hiding and exhibit itself without shame or fear of rejection.

This may also be one reason why oral sex is very high on many men's list of erotic preferences, according to June Reinisch, director emerita of the Kinsey Institute. "There is something related to the woman's acceptance, love, and admiration of a man's penis that he relates to an acceptance of the whole self, that all of him is appreciated and loved." When I was younger, I interpreted a man's desire for oral sex as wanting to be serviced without emotional engagement, and I frequently resented it. As I got older, however, I realized not only that servicing someone (and being serviced) could be a lot of fun but also that it *had* an emotional dimension, and a quite deep one. I began to see it as less about effacing the servicer and more about healing the one who is tended to. The sites where we have been made to feel the most shame are the ones, of course, which most require that healing.

There's something else about the appeal of oral sex. As our culture defines things, missionary intercourse is the most "manly" sexual activity a man can engage in with his penis, because it requires him to be active, performing. Oral sex, in contrast, allows the man the greatest degree of passivity. Just as getting on top may be especially exciting to women by virtue of the gender transgression involved, so getting passively pleasured may bring a special kind of release to men. At the same time and for the same reason—our equation of masculinity with activity—men's delight in oral sex may have a secret shame attached to it. I wonder whether Bill Clinton would have been treated in the press like such a voracious, infantile little boy if his sins with Lewinsky had been of the more active, "manly" variety? I can't help thinking that the "passivity" of his illicit affair—the fact that she was pleasuring him, rather than the other way around—fed into disdain for him. If so, it would be another arena where Bill Clinton played scapegoat for the sexual secrets of his gender.

These attitudes toward oral sex bear, as well, on Clinton's definition of "sexual relations," on which such unmitigated scorn has been heaped. Clinton, as is now well known, claimed in his deposition for the Jones case that he had not had "sexual relations" with Monica Lewinsky—and later justified this to the grand jury by explaining that he didn't consider oral sex to be "sexual relations." Everyone but everyone—including Clinton supporters—mocked him for what was regarded as legalistic hairsplitting. But Clinton's definition may be more generally accepted than that. According to the American Heritage Dictionary, both "adultery" and "sex" refer to sexual *intercourse*. And doesn't anyone remember that all-important line that separated petting and "going all the way"? Female prostitutes charge less for fellatio than they do for intercourse. Teenage girls who have not had intercourse but have performed oral sex still regard themselves as virgins. And American males, as sociologist Debbie Then points out, tend to view oral sex as "a kind of moral freebie." Intercourse *is* privileged in our culture.

One insight that I've gained from reading men's explorations of pornography is that what is going on when women's bodies are depicted in sexualized or aestheticized ways—whether in porn or high-fashion magazines—is *not* adequately described by what feminists have called "the objectification of women." The category of objectification came naturally to feminism because of the continual cultural fetishization of women's bodies and body parts—breasts and legs and butts, for example. But the fact that women's bodies are fetishized does not mean that men find them exciting just because the parts are big, or firm, or slender. Generalization is risky, and I would never attempt to categorize whole genres of porn. But in this respect, heterosexual porn is perhaps significantly different from gay porn. Gay porn, as Mark Simpson points out, often focuses on organs, penetration, in-and-out, without ever showing subjective responses—in facial expressions, for example. The impersonality is a central ingredient of the turn-on. What's going on in the depiction of women's bodies, in contrast, often seems designed to project a certain attitude, or feminine subjectivity. If the woman were simply a collection of body parts, she would have no power to reject, and thus her unconditional acceptance would be meaningless. In order for that acceptance to be experienced, it must be "spoken" in some way. The body must say: "I exist wholly for you. I will never reject you. You *cannot* disappoint me."

Even the fetish of spread legs, for example—arguably the worst of-fender in reducing the woman to the status of mere receptacle—seems to me to use the body to "speak" in this way. If this fetish really re-duced the woman to the status of vaginal receptacle, she would be no more capable of expressing *acceptance* than the piece of liver that Portnoy masturbated with in his parents' bathroom. But spread legs *do* speak. "Here I am," they declare, "utterly available to you, ready to be and do whatever you desire." Many women may not like what this fetish, as I have interpreted it, projects—the woman's willing collaps-ing of her own desire into the male's pleasure. Clearly, my interpreta-tion doesn't make pornography less of a potential concern to feminists. But it situates the problem differently. It's not about the reduction of women to mere bodies, but about what those bodies *express*. Men and women may have very different perspectives on this, which need to be gotten out in the open and disinfected of sin, guilt, and blame.

So too with beauty. Part of the shared knowledge exchanged be-tween me and the woman at the stoplight was recognition that the constant body work women do is not simply about making the body beautiful. It speaks of women's willingness to devote a great deal of time and labor to *making* ourselves up (and even over) for another. The Pretty Women of our culture—that is, the actresses and models who play that role—not only model rather rigid parameters for what counts as beautiful; they also teach women about the deep meaning of another pair of eyes—the gaze—on you. "Here I am," the female bod-ies whisper to us, in painting upon painting, ad upon ad, movie upon movie. "Do you like what you see? If you don't, I can change. What color lipstick would suit? Should I lose a few pounds? Straighten my hair? Speak more softly?"

Men too, their sexual desires honed on movie stars and models, are set up for disappointment—in us and in themselves. Many try out women the way we go through lipsticks, searching for just the right one, the perfect one, the one that will do it. Few will find her, for real women—unlike the images—are not perfect. But the very illusions that *we* force on our imperfect bodies feed men's fantasies and create a bloated idea of our power over them. Just recently, a rash of movies written and directed by a younger generation of men—*Beautiful Girls, Clerks, Chasing Amy, Swingers*—has finally begun to explore the insecurities of "guy culture." Most of them deal, not surprisingly, with the pain of crushes on and rejections by the beautiful women who—as young men experience it—hold all the aces.

Some men have complained of the loss of heroic male imagery in the movies today, finding these films full of characters who are "callous, immature, and generally downsized." I would say, rather, that what's going on is the extremely belated and welcome deflation of a mythology of male stoicism, self-containment, and autonomy. In these films, we get to see men agonizing over relationships: pining and weeping, neurotically analyzing, self-scrutinizing, pulling the covers over their heads, and overeating—just like women. Women viewers who remember sitting anxiously by the phone, wondering what they did wrong (hips too big? voice too loud?), get to see men fretting and stewing with each other over how many days they should wait before calling the girls whose phone numbers they've gotten at the bar. We get to see not just how nervous that phone call can make a man—that motif goes back to Andy Hardy—but how *needy* the man who makes it may be. That's a new motif for popular culture, and it's one that shows that men really *do* come from the same planet as women. We're all earthlings, desperate for love, demolished by rejection.

At a certain point in the semester, my student with the Farrah poster on his bedroom door tore it down. I asked him if he had felt bullied by his girlfriend, by the other women in the class, or by me. Not at all, he said. It was just that he began, at a certain point, to be unable to look at the poster as "just a picture." Farrah still looked beautiful to him, but her beauty now was complicated, cluttered with information and associations that he was not entirely pleased about, but that could not be erased. He could no longer look at Farrah's dazzling smile, or her perky nipples, or her tousled blonde mane without also seeing the hidden artifice behind the California-casual image. Without seeing the power this airbrushed body had over his fantasies, and how bad it made his girlfriend feel about herself. Without thinking of Eldridge Cleaver, hating himself, a pinup of a white girl on his prison wall. Like Cleaver, he explained, he suddenly found himself on the other side of a knowledge that could not be revoked or returned. It wasn't about self-censorship or learning to be politically correct. It was about an increase in knowledge. The knowledge, among other things, that he and his girlfriend weren't shouting at each other across a galactic divide, but merely standing on different sides of the same fun-house mirror, reflecting the distortions of each side back to one another.

humbert and lolita

It took three years for Vladimir Nabokov's *Lolita*, first published in Paris by Olympia Press in 1955, to find an American house willing to publish it. When G. P. Putnam's did so in 1958, some reviewers were baffled by the reputation for prurience that preceded it. Elizabeth Janeway, in her 1958 review for *The New York Times Book Review*, described it as "one of the funniest and one of the saddest books" she had ever read. "Humbert is all of us," she declared. Not everyone agreed. The next day, in the daily *New York Times*, Orville Prescott proclaimed *Lolita* "repulsive . . . highbrow pornography." Forty years and two movie versions of the novel later, our culture is still sharply divided about what to make of Nabokov's story about a middle-aged man's sexual obsession with a twelve-year-old girl.

Nabokov's Humbert Humbert would seem to represent a particularly challenging version of "the male animal" to empathize with. And indeed, I haven't always been able to empathize with Humbert, in part because I hadn't the life experiences to enable me to, in part because my responses to Humbert were influenced by cultural factors. All reading is affected by such influences, but it is dramatically affected in the case of *Lolita*, which continues to irritate the nerves of contemporary sexual politics, and forces us to confront aspects of human behavior that even today we'd prefer not to face. In the late 1990s, incest and child abuse are open topics of discussion, as they were not when

Nabokov wrote his novel. But in the late 1990s, too, we may not be inclined to think of *Lolita* as a story of child abuse. For, in the years since *Lolita* was published, Nabokov's creations have been usurped by cultural archetypes with a life of their own.

For many people, the very word "Lolita" no longer denotes Nabokov's fictional twelve-year-old but exists as an all-purpose signifier for the underage sexual temptress. "Precociously seductive girl" is Webster's definition—and images from the fashion world have given the archetype a visual form. Talk shows, magazine articles, and memoirs describe affairs between older males and younger females—even when the females are not children—as "Lolita-like." In the discussions that follow, the positions have become predictable. On one side are those who view the Lolitas as defenseless victims of male predation. On the other side are those, like Katie Roiphe, who trumpet the "power" younger women hold over their older lovers. On one side, Humbert is a monster—a distinctively male kind of monster. On the other side, he's a pathetic old fool, led by the nose by his baby Eve.

The possibility that "Humbert is all of us" thus remains just as difficult to confront—to really, truly confront, acknowledging all the selfishness, all the cruelty, all the destructiveness of the character, but all his humanness too—as it was in 1958. But this remains the challenge of the novel, for women as well as men, and more so today than ever before.

"how did they ever make a movie of *lolita*?"

The first time I "read" *Lolita*, I skimmed for the dirty passages. I was disappointed to find that they were hard to locate, nestled inside a thicket of fancy ruminations that were as much an obstacle to my pleasure as descriptions of battles, well-appointed rooms, and outdoor scenery were to my enjoyment of the classics we read at school. Like a true girl, I was only interested in "relationships," sexual or otherwise. My standards for sex, in those days, were pretty literal (I got a lot more for my money from D. H. Lawrence than from Nabokov.) Other aspects of Humbert's relationship with Lolita were interesting to me only on those rare pages when Lolita herself had dialogue and action. At the start of the novel Lolita is twelve—roughly my own age when I first skimmed its pages—and there was not yet a genre of romances specially designed for preteen girls. Although the novel begins in 1947 (the year I was born), Lolita—if you left her unorthodox sex life aside—was as close to a realistically drawn, "modern" adolescent as you could find in those days.

Not that I felt a kinship with Lolita, whose coolness was foreign to my personality (Alcott's little women, with their intense passions and ideals, felt more like sisters to me). But still, Lolita was one of the first female characters in whom I could recognize some of the banal details of preteen and teen culture as I knew it. Those moments of recognition made my reading of *Lolita*—a book that had been carefully researched by Nabokov, who rode the school buses in Ithaca, New York, listening to kids chatter, observing their dress and gestures, even recording their average weights and heights—a little like reading an Ann Beattie story today. But as with the sex, I had to endure an awful lot of Humbert's boring, long-winded elaborations, justifications, remembrances in order to get to what, in those days, were *my* pleasures of the text.

The second time I read *Lolita*, it was with Stanley Kubrick's 1962 film version as my guide. At the time, Pauline Kael dripped scorn on critics who didn't like the movie, accusing them of showing off "fake erudition," demonstrating that they had read the book by complaining about the movie's inferiority to it. For me, however—and I imagine for many of my generation—once I had seen the movie, it was *Kubrick's* interpretations that remained etched on my imagination and subsequent understanding of the novel. The movie was the original, the book a palimpsest. And although only a few years had passed since I'd first skimmed its pages, I had become a totally different person. More accurately, I was trying to *make* myself into a totally different person: cool, "avant-garde," satiric, always on the side of the cultural rebel.

The transformation required that my public and private selves be reversed. Now, I carried the transgressive books of the moment—Henry Miller's *Sexus*, Terry Southern's *Candy*—around with me for deliberate effect, and cried over Beth's death in *Little Women* in private, ashamed of my sentimentality. I adored Kubrick's *Lolita* for its scandalous subject matter, iconoclasm, and dark, ironic humor—very much in fashion in those days—and when I went back to reread the novel in 1964, the summer before I went away to college, I either didn't notice or simply didn't care about elements of the novel that Kubrick had left out.

Nabokov, who at first felt only revulsion at "the idea of tampering with my own novel," agreed to write a screenplay for Kubrick's film, and is still listed in the credits. But in fact, little of Nabokov's work made it into the final version of the film—"only ragged odds and ends," as the author commented in the foreword to the published version of his screenplay. Artistically, Kubrick's decision was the right one. I've read Nabokov's screenplay, and concur with his own admission that he was "by nature no dramatist . . . not even a hack scenarist." His screenplay is both bizarre in spots (as Nabokov tried his hand at fancy cinematic effects and juxtapositions) and oddly pedestrian overall, lacking the poetry and psychological precision of the novel. He didn't employ the logical device of a voice-over to convey Humbert's states of mind and preserve the beauty of his own language, but instead created a plethora of new scenes, new dialogue. It is as though, having recognized that the requirements of dramatization would necessarily limit the freedom of interior monologue that is so rich in the book, he decided to go in the opposite direction, and make plot and dialogue the conveyance of all that he wanted to say. Since he still wanted to say quite a lot, the result is an unwieldy, uncentered screenplay.

Nabokov's screenplay also had a big external obstacle to negotiate, as did the Kubrick screenplay that replaced it: the Hollywood Production Code. "How Did They Ever Make a Movie of *Lolita*?" provocatively asked the trailer to the film. Several critics answered: "They didn't." To the extent that they are right, it's due in no small part to Hollywood censorship. Kubrick, in fact, confessed years later: "Had I realized how severe the [censorship] limitations were going to be, I probably wouldn't have made the film." One big challenge was how to represent Humbert's character without the sexual detail of the novel. Throughout much of the novel, Humbert—while discoursing elegantly about childhood loves and nymphet charms—in fact behaves like a lecherous and manipulative scumbag. At the beginning of his relationship with Lolita, he constantly schemes ways to surreptitiously fondle her without her knowing. The night they begin their cross-country journey, he drugs her so he can pleasure himself while she's unconscious. Later in the relationship, he badgers, terrorizes, and bribes her with money into dispensing her sexual favors.

Even suspending judgment on the issue of sex with a minor (not to mention a stepdaughter), Humbert is not a nice man, as he himself recognizes and even boasts of at times, in guilty awe at his own wickedness. Recalling his thoughts about Lolita's inevitable maturation, he remembers plotting that either "I would have to get rid somehow of a difficult adolescent whose magic nymphage had evaporated" or "with patience and luck I might have her produce eventually a nymphet with my blood in her exquisite veins, a Lolita the Second, who would be eight or nine around 1960, when I would still be *dans la force de l'âge.*" He even contemplated the possibility of "Dr. Humbert practicing on supremely lovely Lolita the Third the art of being a granddad." Nabokov means us to be horrified, even as we may laugh at Humbert's speaking so drolly about his planned exploitations of generations of Lolita clones. The horror, I should emphasize, is *not* at the fact that Humbert is in love with a child (an emotional predicament for which Nabokov, as we'll see, has sympathy), but that he can plot her disposal and abuse so callously. The humor is dark—and very fragile. It's easy not to find it so funny.

Obviously, in 1962, Kubrick couldn't have Humbert's narrative voice-over express thoughts like those quoted above or any other musings of an explicitly lascivious nature, or show him continually trying to steal a grope (in the novel, he even "steals a honey of a spasm," that is, has a covert orgasm) while Lolita remains oblivious. (The most Kubrick allows of this nature is Humbert's grab of Lolita's hand during a horror movie, Charlotte sitting right beside them.) Yet Kubrick insightfully recognized that it was crucial that changes made with the censors in mind not work to rehabilitate or soften Humbert in the eyes of the viewer. He knew that Humbert isn't Humbert if there aren't abusive—even monstrous—aspects of his behavior that he (Humbert) recognizes are "objectively" horrible, but nonetheless indulges in and describes with sardonic, urbane wit.

Kubrick solved the problem (as Nabokov did in his own screenplay) very intelligently, by using the relationship between Humbert and Charlotte, rather than Humbert's sexual proclivities, to expose the "hidden" Humbert: the selfish, callous—even cruel—schemer behind the kindly, civilized professor of literature. Reading Charlotte's pathetic written declaration of love for him, he laughs uproariously. (In Nabokov's screenplay, he does a jig.) After Charlotte is conveniently killed by a car, he relaxes contentedly in the bath with a "rapturous

swig of Scotch" (as Nabokov described the detail, one of those added
by Kubrick which the author praised as "appropriate and delightful").
In the film, we know Humbert's not getting drunk out of grief because
of the teasing "ya-ya" music in the background—the Lolita theme—
and because his neighbors, come to express their condolences, are portrayed as clueless fools for thinking he might be suicidal because they
see a gun on the table (he had actually been considering murdering
Charlotte). James Mason's Humbert, who fit to a tee the physical descriptions of Humbert in the novel—dark, ruggedly handsome, but in
a suave European mode—embodied smooth, surface civility and secret, guilty unease to perfection. The division between his public face
(polite, refined, urbane) and his inner life (wriggling, poisonous worms
squirming through his brain) is written on his face in virtually every
scene.

A concern for the censors was a major reason, too, for the choice of
curvaceous teenager Sue Lyon for the title role. In the book, Humbert
is drawn to Lolita's twelve-year-old body, with its "narrow white buttocks" and "puerile hips," its "thin, knobby wrists" and "beautiful
boy-knees" precisely for its immaturity. He's repelled by "that sorry
and dull thing, a handsome woman," and views even the "average
coed" (with her "heavy low-slung pelvis" and "thick calves") as a
"coffin of coarse female flesh." Employing a twelve-year-old actress
was out of the question, however. So, too, was even using an actress
who *looked* like one. We were still pre-Twiggy in 1962, and an undeveloped female body, Kubrick feared, would scream to the Production
Code Administration: "This is a *child* having sex with a middle-aged
man." Sue Lyon's Lolita—her actual age fifteen and her "literary" age
upped to fourteen—had a body that looked like it had been enhanced
by estrogen for some time.

In the novel, Lolita apes the gestures of movie stars. But her still-
boyish, twelve-year-old body wears those efforts like a child. When she
puts on lipstick, it's smudged and overdone. Her hygiene is poor; she
rarely washes her hair and "the only sanitary act she performs with
real zest" is brushing her teeth. Yet Humbert loves her "brown smell"
and dirty jeans. Nabokov, shown Sue Lyon's picture, was reassured by
Kubrick that she "could be easily made to look younger and grubbier"
for the part. He didn't keep his promise. Lyon did bring a bored,
know-it-all smirk to the role, like one long "puh-lease!," that actually

turned her beauty to advantage, suggesting it was something she hadn't yet grown into and giving it an eerie cast, like a preteen version of JonBenet Ramsey painted to be years older than her age for a pageant. But when we first see Lolita in what is arguably the most famous scene in the movie she's a bikinied beach blanket Dolly, all soft curves, sunbathing in the backyard. She knows how to wiggle across a room. Her makeup is expert, her hair Breck-girl coifed. As she sprawls across the bed in the Enchanted Hunters Hotel, where Humbert and she have intercourse for the first time, we see that she is wearing spike heels! Take the gum out of her mouth and she and James Mason are a pretty attractive couple. (The Humbert of the novel, however, would probably have found Lyon too feminine to be desirable.*)

The distortion was exacerbated by the fact that Kubrick chose not to include any of the vignettes from the novel which bring Lolita's misery to the forefront, nudging Humbert's obsession temporarily off center stage. Amidst all the early outrage (and, from different quarters, celebration) over the novel's "prurient" themes and scenes, Nabokov's wife, Vera, insisted—rightly—on "the pathos of Lolita's utter loneliness." Nabokov's Lolita, certainly, is no angel. She taunts and sulks and connives; she has a smutty mouth. But once her mother is gone, she's also an unhappy little captive with "absolutely nowhere else to go" besides Humbert's arms, who squirrels away her allowance in order to collect enough to run away from him. She may talk flippantly to Humbert about her "murdered mummy," but she devours sentimental novels about self-sacrificing mothers and brave little daughters and cries herself to sleep at night—every night. In Kubrick's film, one good sobfest and dead mommy is forgotten. Humbert, to calm her down, has promised her a brand-new hi-fi and all the latest records. The same scene in the novel ends with Lolita sobbing *despite* Humbert's having plied her with gifts all day.

These choices, I believe, cannot be explained away with the "concern for censors" story, but seem to be telling a much older story—the

---

*Pauline Kael defended Kubrick against those critics who remarked on how mature Sue Lyon looked. She suggested they needed to visit a local grade school: "Have the reviewers looked at the schoolgirls of America lately? The classmates of my fourteen-year-old daughter are not merely nubile; some of them look badly used," she wrote. But Kael forgets that when Humbert first sees Lolita in the novel she doesn't look like a "badly used" woman at all, she *looks* like a child of twelve. That's why Humbert falls in love with her.

age-old story—of Eve's responsibility for the fall. If it weren't for
Eve's hand extended, offering Adam the luscious fruit of his own
eventual degradation and sin, we'd all still be in paradise, you know.
This archetype, in the end, claimed Kubrick's Lolita—not so much
for misogynistic reasons as because of the overwhelmingly cynical,
antisentimental character of the movie. Emphasizing Lolita's sadness
and loss would jar with the film's dedication to inflecting the "dark"
with the comic. So all we're permitted to see of Lolita is that which is
manipulative, predatory, and vulgar.

This sensibility reflected very well the times in which the film was
made. In 1962, we were still prefeminist. The avant-garde's "cultural
politics" were directed against fifties conformism, Puritanism, Norman
Rockwell Republicanism, and corn. There was an authentic intersec-
tion here too with Nabokov's own values. His condescending attitudes
toward what he considered to be artistic schmaltz (which included
Dostoevsky) were pretty well known, and—as literary critics loved to
point out—he insisted in his afterword to Lolita that his book had "no
moral in tow." An industry of hip criticism, aided and abetted by the
cynical overlay of Kubrick's film, developed around that supposed
"absence" of morality: the book has nothing to do with incest or child
abuse, we're told; it's about "the logic and delight of games," the death
of meaning, it's a stylish experiment in language.

These critics forgot, however, that Vera Nabokov—to whom the
book and screenplay were dedicated and who lived her life entirely in
the service of Vladimir's art—probably was not merely speaking for
herself in emphasizing the pathos of the novel. Nabokov himself insisted in postpublication interviews that the book had a "high moral
content." I like Nabokov's distinction between towing a moral and
having moral content very much. Lolita doesn't preach or offer a ped-
agogy; it's frequently very funny. Its "moral content" is in the juxtaposition of artistic details. Even as he's drawn us into the thick of
Humbert's obsession, Nabokov encourages us to recognize the disjunction between Humbert's engorged fantasies and the banal, pop-culture
mentality of the child he's screwing, who'd rather be (and should be)
eating a hot fudge sundae. Through perfectly selected details, he reminds us that Lolita is not a "soul-shattering," "deadly little demon"
of "fantastic power," but a child—a real child, an ordinary child. A
child whose stepfather's penis is much, much bigger than anything

adolescent Charlie offered at camp, a child who is convinced—because Humbert tells her so—that she will be sent to reform school if she tells the police about his kidnapping of her, a child whose mother has died. Why doesn't Humbert run to comfort her when she cries? "I always preferred the mental hygiene of noninterference," he explains later. But, "squirming and pleading" with his memory, he is forced to recall, too, that "it was always my habit and method to ignore Lolita's states of mind while comforting my own base self."

portrait of the pedophile as a young boy

Kubrick's film did not adequately represent Humbert's innocence, sadness, or powerlessness either. The Humbert of the novel is not static but undergoes a change. For most of the relationship, the fact that Lolita is not a sexually mature woman but a child moves him to lust. In the end, however, it moves him to sorrow, as he mourns the absence of Lolita's voice from the musical "concord" of children at play (a "spurt of vivid laughter, . . . the crack of a bat, . . . the clatter of a toy wagon") and realizes he had stolen exactly the thing he found beautiful—her childhood—from her. In Kubrick's film, there's none of this painful regret, only torrents of tears at the final loss of Lolita (pregnant and married, she refuses Humbert's entreaties to leave her husband and go with him) and murderous rage at Quilty. Humbert never is shown to realize, as he does in the novel, that while Quilty had "broken Lolita's heart," Humbert was the one "who had broken her life."

The film also omits any reference to the fact that Humbert falls in love at first sight with Lolita because she looks like another child of twelve—named Annabel, and described in the novel with many Poe-like allusions—with whom Humbert, himself then just thirteen, had fallen in love on the Riviera many years before. That child died, and Humbert had been haunted "by her seaside limbs and ardent tongue" ever since. When he sees Lolita kneeling in her mother's garden (kneeling, not sunbathing seductively as in the film), Humbert feels, with "the impact of passionate recognition," that "it was the same child—the same frail, honey-hued shoulders, the same silky supple bare back, the same chestnut head of hair . . . I saw again her lovely indrawn abdomen where my southbound lips had briefly paused; and those puerile hips on which I had kissed the crenulated imprint left by the band of her shorts . . ."

Passages such as this do not give Humbert a moral "excuse" (the last thing Nabokov would want to do) but they do reveal a core of beauty at the center of the poisonous apple, and—along with Plato—suggest the essentially nostalgic nature of all love. The mere sight of a face, sound of a voice, glimpse of a body poised in a certain way, can indeed collapse the passage of time, recalling memories of our former selves before (as Plato might put it) we fell to earth, out of the realm of the ideal and into the world of the actual. The three times in my life that I have fallen in love "at first sight," this happened; some combination of features, movements, gestures too unpredictable to fix, ineffable but utterly individualized (the three men looked nothing like each other) instantaneously claimed an already existing imprint on my heart, unknown to me until that very moment (unlike Humbert) but (like Humbert) echoing back to a time when I still believed in dreams—of bliss, acceptance, a perfect meeting of souls and bodies—that mattered deeply to me. Humbert's dreams, like mine, were innocent; it is only in the realm of action, never in the realm of the imagination, that such dreams are anything but innocent.

Why did Kubrick leave out the Annabel memory? Again, I suspect that he did not want an elegiac mood to interfere with the ironic, comic tone of his movie. The Annabel incident is, however, a not insignificant detail to have left out, and so is the theme of remembrance of things past. In his memoir *Speak, Memory*, Nabokov confesses that "I do not believe in time. I like to fold my magic carpet, after use, in such a way as to superimpose one part of the pattern on another." It is an essential element of Nabokov's approach to life, as well as art, that in the presence of certain faces (or sunsets or butterflies or material objects) that carpet unfolds and meaning is released. *Speak, Memory* and *Lolita* are full of such unfoldings. After Lolita has run away, Humbert cleans out his car: "My Lolita! There was still a three-year-old bobby pin in the depths of the glove compartment. There was still that stream of pale moths siphoned out of the night by my headlights." It's a perfect, small poem about the irrevocable pastness and continuing presence of the imprints of intense love.

By giving Humbert his own artist's sensibilities, Nabokov shows us that Humbert is not just a charming monster with a thing for little

"Will the real Humbert Humbert please stand up?": James Mason sneaks a lecherous peek at his stepdaughter (Sue Lyon) in Stanley Kubrick's *Lolita*; Jeremy Irons as "nice" Humbert, off on a trip with his girl (Dominique Swain) in Adrian Lyne's version

girls, but is haunted by something much deeper than mere lust, or even personal memory. He's also a relic of a less dispose-all culture, tormented by what might be described as an exquisite hypersensitivity to beauty, finding it even in the most banal of objects, excruciatingly pained at its passing, unable to let it go. Humbert may joke about disposing of Lolita when she's past her nymphage. But actually, he's a man who can't even throw her bobby pin away. When he finally does see her again, pregnant and married and "only the faint violet whiff and dead leaf echo" of her nymphet self, he realizes that he still loves her: "No matter, even if those eyes of her would fade to myopic fish, and her nipples swell and crack, and her lovely young velvety delicate delta be tainted and torn—even then I would go mad with tenderness at the mere sight of your dear wan face, at the mere sound of your raucous young voice, my Lolita."

It's this sad, memory-haunted, backward-looking Humbert that Adrian Lyne's *Lolita* gives us—in spades—erring in exactly the opposite direction from Kubrick's film. It's as though Kubrick chose to present us with one half of the character and Lyne with the other. Each gives us a "Humbert" but neither gives us "Humbert Humbert." The Humbert each chooses to give us—even more, the Humbert each chooses to censor—reveals more about the cultural zeitgeist of each director's own times than they do about Nabokov's character.

From the opening credits of Lyne's film, the film announces an interpretation that is virtually the mirror image of Kubrick's. Kubrick's opening credits were both transgressive and ironic: a close-up of Lolita's foot as Humbert (unseen except for his hand) painstakingly paints her toenails with polish, while Nelson Riddle's ersatz Rachmaninoff piano music wrings its heart in the background. Kubrick's tongue is in his cheek from the start. Lyne's movie, in contrast, begins (and pretty much remains) in nostalgic, dreamy soft focus, as Humbert, having just murdered Quilty, eyes brimming with tears, virtually in a swoon, drives his car down a country road. Mist is everywhere, both visually and musically, as Ennio Moricone's minimalist, elegiac, lyrical score (no overblown romantic crescendos in Lyne's film) plunks little teardrops of notes here and there. A voice-over skips the first paragraph of the novel ("Lo-lee-ta: the tip of the tongue taking a trip of three steps") to begin on the more nostalgic, tender note of the second: "She was Lo, plain Lo in the morning, standing four feet ten in one sock . . ."

"Did she have a precursor?" In Lyne's film, yes. The actress looks to be about eighteen, and straight off the pages of *Glamour* magazine, complete with fashionable straw hat and pretty camisole, which she unbuttons seductively in front of a watching Humbert. In cinematography, the episode is even more drenched in Old World vapors than the rest of the movie. The relationship, however—unlike the clumsy, fevered, mutual groping that occurs in the book between twelve-year-old Annabel and thirteen-year-old Humbert—is something from *Showtime*'s late-night *Red Shoe Diaries* (minus the complete undress and coupling, of course). It's a boy's fantasy—not a boy's memory—of a first sexual encounter. As soon as I saw Annabel's breast curves, her languorous body as she stretches on the beach, her sultry glances from beneath lidded eyes, I knew that something fatally wrong was going to happen in this movie.

Much praise has been lavished by critics on screenwriter Steven Schiff's liberal use of exact quotes from the novel in the film's voiceover. Nabokov's narrative of the Annabel episode was apparently not enough, however, to make the exculpatory point that Schiff wanted, so he supplied some very non-Nabokovian language of his own to drive home a psychological "motivation" for Humbert's later behavior. "Whatever happens to a boy the summer he is fourteen can mark him for life," we are told portentously at the outset of the Annabel episode. (Yes, Lyne decides to make Annabel fourteen too—even though public morality would presumably find nothing shocking about a twelve-year-old and a thirteen-year-old necking in the sand.) Then, at the end: "The shock of [Annabel's] death froze something in me. The child I loved was gone, but I kept looking for her long after I'd left my own childhood. The poison was in the wound, you see, and the wound wouldn't heal." This contemporary theory of Humbert as a wounded child, passively suffering the poison of unhealed hurts, is the film's dominant understanding of him—and suffuses Jeremy Irons's portrayal, which has nothing wicked, scheming, or even mildly lecherous about it.

Irons plays Humbert as a man who's simply tragically in love with the wrong woman, and doesn't know what to do about it. Around her, he's not only haunted by boyhood memories, he's as shy, awkward, and tentative as a boy himself, gazing at her like a lovestruck puppy, and never, ever trying one illicit fondle (never even contemplating one) until she's "seduced" him at the Enchanted Hunters. (Lyne, like Kubrick, conveniently omits Humbert's drugging of Lolita the night before.) Unlike Nabokov's Humbert, Lyne's Humbert wouldn't dream of groping Lolita without her consent.

Lyne, Schiff, and Irons all stress, in a documentary called *The Lolita Story*, that they felt it important that the audience *like* Humbert, that they find him (as Schiff put it) "appealing, although he is a monster." I agree with the theory. But in fact there's nothing monsterlike in the slightest about the Humbert they've created. Frank Langella's truly slimy Quilty, robe flapping, stroking his own naked thigh as he promises Humbert "unique erotica" if he spares his life, absorbs all the monster plasma in Lyne's *Lolita*. Unlike Kubrick's Humbert, Lyne's doesn't even have much of a sleazy side with Charlotte. When Melanie Griffith screams the words to Humbert—"You're a monster!"—it seems excessive, although Humbert's vile diary is in her hand. On his part, Irons just stands there like a helpless, trapped little boy (in the same scene in the Kubrick film, you could see James Mason's mind turning, trying to devise a clever way out even as Charlotte shrieked).

lola and hum

Quilty is the film's sole monster. And little Lolita, first appearances to the contrary, is the film's sole sexual seducer, manipulator, schemer. Fourteen-year-old Dominique Swain was a controversial choice for the title role. She doesn't look twelve (and Lyne, following Kubrick, has aged the character to fourteen), but she does have a leggy, lithe body that is more faithful to the Lolita of the novel. Presumably to make her look even more like a child, Lyne has loaded her with accessories ostentatiously advertising her girlhood: old-fashioned plaits wrapped around her head, retainers on the teeth, milk mustaches, oversize pajamas. Ladies and gentlemen of the jury, don't be seduced by this artfulness! Lyne's Lolita is a child only when she's guzzling milk in front of an open refrigerator, or crying into her pillow at night, or rolling a jawbreaker around her mouth. When she's after sex or money, she's all woman.

In one scene, she's sitting across the room from Humbert, trying to convince him to let her be in the school play. A knowing look passes over her face, and she crawls over to where he is sitting in a rocker and begins rocking the chair with her foot. "I have a right to be in a play," she pouts. "Not if I say you don't," he replies. She moves her foot to caress his inner thigh. "You like that. . . . You want more, don't you? Well, I want things, too . . ." She expertly moves her hand up his thigh, her fingers inching toward his crotch. "I think my allowance should be two dollars . . ." (Still closer to his groin; by now, she had him virtually in the palm of her hand, so to speak. "God, yes," Humbert agrees.) ". . . and I get to be in the play."

Lyne's Lolita is a highly passionate girl too and—although she's deeply wounded and cries like a baby when Humbert smacks her across the face in anger—appears to get turned on by a little sexual violence. In one scene, Humbert rapes her, inflamed by the sight of her, her lips smeared with scarlet, apparently having returned from a date with another lover. Lolita had tried at first to lie, but when she realizes he's not buying it, she gives up playing the innocent and smiles provocatively, a bag of bananas that Humbert's just bought her dangling from her hand, in a quintessential "she's asking for it" moment. As he throws her on the bed and enters her, she becomes sexually excited, giving him ardent, deep tongue kisses, then alternately cackling wildly and swooning with pleasure at his rough treatment.

Viewers unfamiliar with Nabokov's masterpiece (or whose memory of the book has been dimmed by time) are unlikely to recognize that Nabokov's Lolita never behaves anything like this. Yes, the novel's Lolita is careless, experimental with her body. Humbert was "not even her first lover." But sex with adolescent Charlie (at summer camp) had "hardly roused" her; she thought it was a kid's game, "part of a youngster's furtive world, unknown to adults." Lolita had done the deed, but her sexual consciousness was still that of a child. And yes, in the book Lolita does barter sex for money. But it's always Humbert who initiates things, Humbert who's doing the begging, the pleading. Lolita, in Nabokov's words, merely exploits "her power to deny" him. This "drop in Lolita's moral" (as Nabokov, oozing irony, has Humbert describe it) occurs only after she has been with Humbert for some time, continually badgered by him for her favors.

As far as Lolita's own sexual passion goes, in the novel it never even takes off, let alone reaches the heights we see in Lyne's film. In fact, there's no evidence that she is sexually responsive at all. Humbert describes her as his "frigid princess," who invariably "preferred a hamburger to a Humberger," and at best would tolerate his caresses when promised money or presents. "Never did she vibrate under my touch," he recalls. For Humbert, who is revolted by the "pitifully ardent" passion of mature women, Lolita's sexual diffidence is a large part of her appeal. "There she would be," he recalls fondly, "a typical kid picking her nose while engrossed in the lighter sections of a newspaper, as indifferent to my ecstasy as if it were something she had sat upon, a shoe, a doll, the handle of a tennis racket, and was too indolent to remove." Lyne and Schiff take *this* image of Lolita reading the funny papers, oblivious to Humbert's mounting passion, and convert it into an atmospheric, hazy, sun-streaked romantic idyll in which Lolita, sitting on Humbert's lap, is overcome with sexual excitement and appears to have an orgasm.

Kubrick's physical maturation of Lolita, as we've seen, was to appease the censors. An orgasmic Lolita, in contrast, seems to be deliberately upping the sexual ante, in defiance both of the novel and of some notion of public morality or sexual politics. Did Lyne believe it was his mission to correct some imbalanced view of parent-child incest? To show us that stepfathers who sleep with their daughters may actually be giving them a pretty good time some of the time? Did he believe we needed to be educated about the sexuality of children, that they have orgasms too? When Lyne's film was spurned by U.S. distributors for nearly a year while it played in various cities in Europe, he complained of the tyrannies of political correctness in America. It's fair to ask: just what "politically correct" views was he trying to challenge in his film?

In this regard, the comments that both he and Schiff have made about their conception of the Humbert/Lolita relationship are illuminating. Lyne has said he wanted to portray an "evenly weighted battle of the sexes," Schiff that he viewed Humbert and Lo as "a couple—a very odd couple, to be sure, but a couple nonetheless . . . testing each other, confounding each other, and, yes, loving each other." He certainly succeeded in confounding Michael Wood, who wrote in *The New York Review of Books* that the orgasm scene in the movie "seems relatively healthy, since it's only sex, and both are enjoying themselves. I know it isn't healthy, and I don't think it is once I remember her age. But I have to make myself remember this."

He wouldn't have this difficulty if he went back to read the book. The idea of Hum and Lo mutually pleasuring each other (*The Joy of Sex*, perhaps, on the bookshelf behind them) is utterly at odds with Nabokov's depiction of the relationship, which (as I hope my quotations make clear) is about as far from a May-December romance as can be imagined. But perhaps we shouldn't be surprised that Adrian Lyne, as much a Geiger counter of our contemporary sexual zeitgeist as Stanley Kubrick was of the sixties, has created such a sexually knowing, at times predatory Lolita. This Lo's sexy sisters slouch all over the fashion magazines, are celebrated in "power feminist" memoirs like Katie Roiphe's. Roiphe was sixteen—hardly a child—when she had an affair with her thirty-six-year-old teacher. Yet *Vogue* advertised her piece with the gloss that "*Lolita*-like affairs are not always about a predatory man and a victimized girl. For Katie Roiphe, the relationship was an unusual balance of power."

Here's her "power," as described by Roiphe: "He felt guilty. I had something he wanted . . . He and I bargained over each act of physical intimacy like hagglers in an open market, and in the course of that bargain, I learned a kind of sexual control that would stay with me for the rest of my life." What a sad, awful conception of female power and control . . . and a very old one. Using a fourteen-year-old actress in graphic sex scenes may cross the line with the Child Pornography Prevention Act of 1996 (hence, the love scenes in Lyne's *Lolita* have far less skin than we're used to, and a double had to be used when certain acts were depicted). But images of women as the mighty controllers of male sexuality, holding that all-powerful "yes" or "no" switch, have an ancient and enduring lineage in our culture. After all, it wasn't that long ago that some lawyers were still defending child molesters on the grounds that the little girls had seduced them.

Precociously seductive girl. Webster's definition for "Lolita," you'll recall. In the documentary *The Lolita Story*, Lyne even describes the novel itself in such terms, speaking in honey tones of its continuing "capacity to charm, to shock, to break the heart of a man . . . and a filmmaker." He's hardly the only one who's got his Lolitas—character, novel, cultural archetype—all mushed up. After his film was screened for a preview audience, one woman said to screenwriter Stephen Schiff: "I guess you called her Lolita because she was really kind of a Lolita, huh?" Rough translation: "She was a sexy little slut, wasn't she?" No, she wasn't, not in the novel. Her "innocent mouth" tries out the romantic gestures of her movie star idols. Her "guileless limbs" are carelessly exhibitionistic. But she never lasciviously creeps across a floor, expert fingers making their way to Humbert's crotch. She never has an orgasm on Humbert's lap. She takes no pleasure in being raped.

If Lyne were truly interested in being as "honest" to Nabokov as he claims he tried to be, he would have worked against the cultural archetype. Instead he decided to dress it up in gingham, making period details and carefully chosen, lyrical quotations pass for "fidelity" to the novel. Judging from the critics' reaction, he seems to have pulled it off. Almost without exception, the reviewers, who've presumably read the novel, have remarked on Lyne/Schiff's faithful rendition: a "lavishly faithful production" *(The New York Times)*, "almost debilitatingly loyal to Nabokov's novel" *(The New York Review of Books)*, "careful, even loving, loyalty to the book" *(The New Yorker)*.

"The film's master stroke," Caryn James writes in *The New York Times*, "is its understanding that this is Humbert's story, told in his own lyrical voice, from his own passionate, sad, tortured perspective." Well, no, not quite. That's the film's master *error*. Vladimir Nabokov wrote the book, not Humbert Humbert. It is Vladimir Nabokov who has given Humbert a voice, Nabokov who constructs the architecture of the whole and fits the details he's chosen within it. When it comes to Humbert's character, those details, as we've seen, are creepy and crawly as well as lyrical and sad. And when it comes to Lolita, Nabokov, although sympathetic to Humbert's memory-driven obsession, refuses to allow the reader to share Humbert's mythical view of her sexual power. Lyne, in contrast, gives us lyrical Humbert but not lecherous Humbert, and a Lolita who behaves with him like a junior Monica Lewinsky, exulting in the power of her fingers and tongue to keep the horny old guy coming back for more. There's pathos in that story, too—but it's not the pathos of *Lolita*.

a culture of humberts and lolitas

Humbert is all of us. I return to Elizabeth Janeway's provocative declaration. Humbert, Janeway goes on, underlines "the essential, ineffi-
cient, painstaking and pain-giving selfishness of all passion, the greed
of all urges, whatever they may be, that insist on being satisfied without regard to the effect their satisfaction has upon the outside world."
I think she nailed it pretty accurately, and helps us to see where both
Kubrick and Lyne missed the mark. Kubrick's *Lolita*, in emphasizing
the wickedness of Humbert's obsession without revealing the sadness
at its root, reduced the possibility that we might see our own longings
and vulnerabilities in him, that we might recognize that "Humbert is
all of us." But Lyne makes it all too *easy* to like his dreamy, passive
Humbert, sparing us recognition of the truly awful (not merely tragically misguided) "pain-giving selfishness" that Humbert—and by ex-
tension, all of us when operating in a greedy "Humbert mode"—are
capable of.

Jeremy Irons put it tellingly when he said that Humbert's character
demonstrates that "nice people do terrible things." Humbert Humbert—a "nice person"? Tormented, yes. Brilliant, yes. Ultimately filled
with heart-wrenching regret, yes. But nice? Nabokov, I believe, wants
us to achieve identification with Humbert less cheaply, not by realizing
that he's "nice people" (just like the rest of us), but by confronting the
fact that the rest of us aren't so nice either. Maybe an Englishman can
imagine niceness as a possibility, even a desirability, for human nature.
But for a Russian like Nabokov (it's one area where he would be in ac-
cord, I think, with Dostoevsky) "nice" just doesn't cut it. We don't get
to the depths of the human soul, its capacity for good *or* evil, through
"nice." *Seeing*—truly, honestly, seeing—requires sacrificing the illusion
that any of us are "nice."

For all of (what passes for) our sexual "openness," we are generally
unwilling to face what Nabokov, with much more integrity, tried to
make so palpably real: how easy it is for human desire to become selfish, abusive, murderous of the very things it finds beautiful. Our
censors won't let certain representations through. But those censors
aren't always who we think they are and the things that are forbidden are sometimes different from what we imagine them to be. The

film's infidelity to the novel aside, Lolita's orgasm scene may seem bold and sexually transgressive. But it's possible—I would say, necessary—to flip things around and ask what Lolita's orgasm obscures, even conceals. It's harder, as Lyne and Schiff surely understood as they tried to adapt Nabokov's details to their May–December conception of the relationship, to see an orgasmic Lolita, swooning on her step-dad's lap, as a less than willing and equal partner in the relationship. By the same token, it's harder, too, to think of Humbert as a selfish predator, taking advantage of the power difference between them. At the very least, he's a pleasure-giving predator, a politically correct pedophile.

In the novel, in fact, Lolita's sexual pleasure matters to Humbert no more than Cindy Crawford's pleasure matters to the adolescent boy masturbating to her image in a magazine. Yes, Humbert's in love, obsessed, he must have her, he's seething, sex-mad, beseeching, out of his mind. But the fact that Lolita has no real interest in him provides not a cool drop to quench the flame. She barely exists—as a sexual subject—for Humbert. The way she *does* exist for him, as he himself puts it, is as his "own creation," his fantasy-nymphet, whose reality he continually "solipsizes" in favor of his own imaginative construction. That's the only way he can feel free to molest her. He's a civilized man, after all, and knows that one doesn't have sex with children. But with Lolita "safely solipsized" (while on his lap, let's not forget) and Humbert carefully modulating his behavior so that his mounting excitement, and his erection, remain concealed from her, the stiflingly domestic suburban den vanishes, to be replaced by a "self-made seraglio," in which the beast in man can release itself without fear of "ridicule or retribution," where he is no longer "Humbert the Hound, the sad-eyed degenerate cur clasping the boot that would presently kick him away," but "a radiant and robust Turk, deliberately, in the full consciousness of his freedom, postponing the moment of actually enjoying the youngest and frailest of his slaves."

Solipsize. Humbert's own word for his behavior toward Lolita is stark and ugly, not at all "nice." Yet when sexual desire is honed on fetishized images—and a man doesn't have to be a pedophile to know what *that* education is like—the need to solipsize the real person in order to keep desire flowing is not the special province of the monster. A friend of mine remarked to me, in discussion about *Lolita*, that he had always felt like a dirty old man, even when he was an adolescent. I didn't ask what he meant by that, because, having done all my research for this book, I think I know. The secret masturbations, glossy magazines in hand, behind closed bathroom doors, or with flashlights in bed. The "intrusive images," firm tits and ass taunting from catalogues mailed to women, pretending innocence. Eldridge Cleaver, sick to death with his lust over the picture of the woman who led to Emmett Till's death. The man who hopes Viagra will make his real wife, with her real flesh, desirable to him again. Pert little girls on the football field, jumping in the air, panties flashing. They're so pretty, aren't they, when they're young and bubbly and unaware. But Lyne is into rehabilitating Humbert in our eyes, not confronting the secrets that Everyman shares with him.

Nabokov, living in a very different era and not much inclined to cultural criticism, was not interested in exploring Humbert's psyche on a sociological level. But in 1998 Janeway's assessment that "Humbert is all of us" resonates culturally in even more specific ways than either Nabokov or Janeway could have predicted. The dirty secret of our culture is hardly childhood sexuality. Freud took care of exposing that one, I believe. It's the other secret, the one that he may have covered up that we're still concealing. I'm speaking, of course, of our eroticization of children. It's far more pervasive now than it was when Nabokov wrote his novel. Then it could be imagined that a taste for the waif was the province of a special sort of pedophile. Today one doesn't have to loiter around schoolyards in order to catch an illicit glimpse of a twelve-year-old, her body in an unknowingly revealing or sexually piquant posture. Just open a fashion magazine. The models themselves may not actually be minors, but the careless poses they strike with their skinny, long-limbed bodies could be illustrations for a coffee-table edition of *Lolita*.

But we don't want to confront the fact that we have become a culture of soft-core Humberts. Far harder for us to digest than images of "precociously seductive" girls are images that *don't* erase the difference in power between adult and child. Recall, in 1995, when Calvin Klein was forced, submitting to public protest, to yank a series of CK perfume ads from television and magazines. In these ads, the waif-like bodies of young boys and girls were arranged awkwardly, underwear exposed, as though they weren't posing for a picture, but caught un-

awares. In the television version of the ads, the voice of the adult pho-
tographer could be heard telling the child to take off her jacket, turn
this way or that, while she performed for him guilelessly, awkwardly.
People were unnerved by the presence of the adult, intruding into the
world of the child, manipulating and arranging her body—exposing
her—while she remained oblivious.

In the usual fashion ad—unlike the censored Calvin Klein ads and
unlike Nabokov's *Lolita* but *like* Lyne's—the adult's lascivious, intrusive gaze is obscured rather than acknowledged. So we can all pretend
that there's nothing "not nice" going on. The fashion designers can
think *they're* nice people because they don't hire minors. The parents
of children who participate in kiddy beauty pageants can imagine
everything is fine if their little girls claim to enjoy the pageants (while
seductive photos of JonBenet Ramsey continue to be plastered everywhere, even by commentators—like Geraldo Rivera—who have declared them "obscene"). And Adrian Lyne can boast to *Premiere*
magazine about what a thoughtful man he was, protecting Dominique
Swain by putting a pillow between her body and Jeremy Irons. A moment later, he turned to the interviewer, his face alight with excitement
over the scene: "It's sexy, isn't it?"

Well, it *is* sexy. But when was that ever the point of *Lolita*? When I
was twelve. And then when Adrian Lyne decided to make a film of it.

The third time I read *Lolita*, I was eight years older than Humbert
Humbert was when he died of a heart attack in prison, at the age of
forty-two, awaiting his trial for the murder of Clare Quilty. Reading
the novel this time, I found myself, like Vera Nabokov, intensely
moved by Lolita's plight. As I read certain passages aloud to my husband, I burst into tears that didn't want to end. My tears, I realized,
were not only for the Lolita I was now in a position to recognize—the
desperately unhappy child trying to make the most of a horrible situation—but for the fact that I, like Humbert, had once callously ignored
her. My tears were for all of Lolita's descendants: the seemingly confident and put-together young girls who had been in my classes and
come to my office, often to cry *their* eyes out, in release of the secret
pain—the eating disorders, the depression, the self-cutting, the abusive
relationships—our public view of them ignores. And they were for the

me, too, that I had ignored, trying to be so hip and unsentimental when I was a teenager in the sixties.

By the time of my third reading, I had spent at least a decade of my life researching and writing about young women's feelings of shame about their bodies, needs, hungers, and about the cultural images that inspire that shame. I was struck by Nabokov's detailed descriptions of Lolita's prepubescent body. Nabokov was an artist, not a sociologist. But with his keen eye and ear for detail, he now seems uncannily prescient in giving Humbert a taste for undeveloped, coltish beauty—and a disgust with mature womanliness—that's hardly an eccentricity anymore. I thought of the students who had written in their journals of their disgust—very much like Humbert's—at their own bodies plumping out into womanhood, of their admiration for the casual, desireless sex appeal of models like Kate Moss. I thought of my own losing struggle to become the slender, cool icon that I believed men desired, my shame over being "too much"—consciousness, will, passion, flesh—for them, my conviction that no matter how I trained and tamed and trimmed myself, I would always be, like Charlotte Haze, the monster of the story.

The third time I read *Lolita*, I was less titillated and amused, more emotionally moved by the novel than I had been when I was seventeen, reinventing myself the summer before going away to college, determined not to let anything nostalgic or sentimental into my being. By then, I knew what it was like to have my own heart twist at the sight of a beautiful body half my age, to try to stop time in its embrace, to be caught up in the unexpected revival of old dreams and past selves, suddenly speaking as urgently as they had when I was a girl. I had learned how unrealistic it is to expect another person to act as the means of one's own renewal, yet how hard it is to give up that desire when everything in one's own emotional history—like everything in Humbert's—is telling you that this is the one with whom the script can be rewritten, the death can be undone. By the time I read *Lolita* for the third time, I had been mistreated both by men and by my own fantasies, and I had mistreated both of them, too. My father had died, and in my grief—so much deeper, longer, purer than I had ever expected it to be!—my love of men, my sympathy for them, my identification with them, was somehow released from the prison of old angers and wounds.

By the time I read *Lolita* for the third time, I was no longer a hipster, had made friends again with the little girl who cried over *Little Women*. Nabokov himself might have disdained such tears as sentimental. But whatever his "attitudes," Nabokov's writing is full of details that are compacted with emotion, and this time around with the novel, his magic-carpet unfoldings of bodies and things made my own memories speak. The "impact of passionate recognition"—in a college elevator, not a garden—and the certainty that the tall young man in the frayed khaki coat, a stranger until that moment, had always been and from that day always would be in my life. Pistachio shells left in an ashtray for two years, remnants of a tumultuous cross-country trip with a less permanent love. The day I decided to sell the car, I realized that they were the reason I was on my way to the dealer; I couldn't clean out the ashtray, so I had to get rid of the car. The frail shoulders of my aging cat, bones poking out where once there had been solid, vital, animal flesh: I cupped them in my hands, and saw my father's spirit in my old cat's eyes, felt my father's dying body alive again under my touch. My cat was warm and happy, meatloaf-style on my stomach, his dear diminished body gently rising and falling with each of my own deep breaths. I looked in his eyes again. My father was there. I am certain of it.

coda

MY FATHER THE FEMINIST

Most of the images of Judaism that I grew up with were strongly patriarchal. Some of these images came from popular culture: Charlton Heston on the mountaintop, the immutable Commandments in his arms. Some of them came from Christian renderings of Judaism, to which all Jews are subjected in this culture: the stern, unforgiving God of the Old Testament versus Jesus, God of Love. Some of them came from my slender understanding of the (undeniably) patriarchal laws and institutions of Judaism. But most of them derived from childish analogy with my father; he was a stubborn patriarch and he was Jewish, so Judaism must be stubbornly patriarchal too.

My father made most important decisions for our family: where we would live, how we would spend our money, what we would do on the weekends. In *Manhood in the Making*, anthropologist David Gilmore describes Jewish-American culture as "one of the few in which women virtually dominate men." This information would have come as a big surprise to my mother. My father's preferences shaped our lives. Because he didn't like my mother's relatives, we rarely saw them. Because he didn't swim or enjoy the beach, our trips to the Jersey shore (essential for deliverance from the sweltering heat of the Newark tenement in which we lived) were for walking the boardwalk, loading our stomachs with hot dogs and knishes, and playing Pokerino. Because my father adored the racetrack, we spent our family vacation at a shabby boardinghouse in Saratoga Springs, New York. At the track in the afternoon, we kids would pick through the sad, discarded tickets which littered the pavement, looking for carelessly castaway "place" or "show" winners, restlessly marking time until he had placed his final bets and we could leave.

My father's Jewishness was rent with contradictions. Brought up on Hollywood and hot dogs, he regarded the religiosity of my mother's family as an Old World encumbrance, and—as I've said—he made the rules. We did not belong to a temple, did not attend Hebrew school, and although we did not have a tree (my father drew the line at what he took to be goyish paraphernalia, symbols of Christianity), exchanged our presents on Christmas morning. My father regarded his WASP brokers and their blonde wives as classy and my mother's relations as "peasants"; but he loved taking the family on a weekend afternoon to the Brownsville section of Brooklyn, which even into the fifties was an Old World Jewish ghetto: pushcarts, live chickens, Yiddish theater. Loss invariably brought out a vocabulary of ritual deeply planted in his body: on Yom Kippur he davened in Yiddish; at my mother's funeral, he tore at his shirt (an Old World symbol of grief).

Like Job, my father was continually in conversation with Jehovah about the trials and sorrows of his life. His sense of personal injustice was cosmic. In *Avalon*, Barry Levinson's film memoir of growing up in an immigrant Jewish family, Levinson presents an incident that nearly jolted me out of my seat in recognition. The eldest brother of the clan—at this point in the film he is about seventy years old—arrives for Thanksgiving dinner to find that his younger brother has already carved the turkey. It had been the unquestioned family custom for the assembled group to wait until the eldest brother's arrival, in a yearly proof of respect for his authority. This time, screaming kids and growling stomachs have won out over symbolic fealties to the patriarch. The eldest brother, injured beyond repair, swears never to speak to his younger brother again. *And doesn't.*

When the needs of others collided with and overruled my father's expectations of how things should be, his overwhelming tendency, like the eldest brother in *Avalon*, was to interpret this as lack of respect; his retaliation, although rarely physical, could be as harsh, irrevocable, and consequential as Zeus's. My half brother, my father's son by an earlier marriage, made the fatal mistake of siding with his own wife "against" my father in an ugly (and, to me, still obscure) exchange of insults and injuries. I know that my father's generosity—a badge of pride with him—had been called into question and that terrible words were spoken and written, words that cut my father to the quick. It was a genuine crisis, an awful fight. But when my father declared that Gary was henceforth cut out of his life, I could hardly take that threat seriously. I was wrong. My father died a decade later without ever having spoken again to his son.

Traditional Judaism *is* patriarchal. It would be an anomaly within dominant Western cultural traditions if it were not. But there is an element within Jewish paternalism that seeds its own deconstruction. I found this amazing story "from the time of the Romans" in a book about Bar Mitzvah—the Jewish ceremony that marks when a boy becomes a man:

> A group of rabbis were arguing an issue of religious law. Rabbi Eliezer ben Hyrcanus offered an opinion. It was rejected. Rabbi Eliezer protested the decision. "I am right and I can prove it. If my opinion is correct, let the stream outside this study house flow backward."
>
> The stream began to flow backward.
>
> Rabbi Joshua ben Hananiah, who led the majority, said, "A stream doesn't prove anything."
>
> Rabbi Eliezer continued, "If my opinion is correct, let the walls of the study house prove it."
>
> The walls started leaning toward them.
>
> Rabbi Joshua held firm, and told the walls to go back to their place.
>
> Finally, Rabbi Eliezer said, "Let Heaven itself bear witness that my opinion is the correct one."
>
> A voice came from out of the sky. "Why do you reject Rabbi Eliezer's opinion? He is right in every case."
>
> To which Rabbi Joshua responded, "The Torah is not in Heaven. We pay no attention to voices."

Pay no attention to *God's* voice? I read this story aloud to my (Protestant) husband and his mouth fell open. But then, so did mine. The association of Jewish scholarship with radical and ongoing disputation and debate is not new to me, of course. I have no schooling in Judaism, but I know that Jews consider the Torah—God's covenant with Israel, given to Moses at Mount Sinai—a much more extensive and flexible body of literature than the classic pictorial version of Ten

Commandments indelibly etched in stone. Far from ten straightforward commandments, the Torah includes not only "all the rules, obligations, history, poetry, and literature contained in the first five books of the Bible" but also "all subsequent interpretations and adaptations." The latter, known as the Talmud, is the subject of continuous interpretation and contestation among scholars and theologians. Within popular culture too, the Jew's enjoyment of and limitless capacity for tenacious argumentation has been a motif of many Yiddish jokes and humorous self-depictions of Jewish life. In *Avalon*, the family engages in a running debate about the most efficient route from New York to Brooklyn, jousting with each other in the same combative, stubborn style as Rabbi Eliezer and Rabbi Joshua. The implication: if we don't have something serious to debate, we'll invent a dispute around something trivial, just to satisfy the itch for argument.

The Eliezer/Joshua story, however, is not about how Jews love to argue. It's a parable about where to look for truth. In the narrative, a contest is staged, ostensibly between two competing human interpretations (Rabbi Eliezer's and Rabbi Joshua's), but *actually* between human argumentation and Immutable Law as two "genres" of knowledge. The dramatic way in which this contest is depicted is extraordinary. First, the physical world "confirms" Rabbi Eliezer's opinion, producing miraculous occurrences (the stream flows backward, the walls lean) that we *expect* to function in the story as proofs of the authority of the one who has called them up. Readers of the Bible are used to the sea parting for the righteous, not for the errant or the erring. So Eliezer seems positioned, by narrative convention, to be the ultimate victor of the debate. This victory seems inevitable when God himself comes out on Eliezer's side. But in this story, remarkably, even God's opinion is "put in its place," subordinated to the authority of human reason. Rabbi Joshua has the final word. It is by adjudicating scholarly conversation on its own terms—that is, by evaluating the rigor and validity of the respective arguments, *not* by seeking external validation from God—that we decide which opinion is right.

Nothing is written in stone. Not even, presumably, the authority of patriarchal laws and institutions. Come up with a good argument against them, and . . . we'll see. This is indeed what I grew up believing. So apparently, too, did many of the young Jewish women who formed the core of "second wave" feminist politics in the late sixties and early seventies. We thought we had an unanswerable argument, its logic evident, its claims for justice compelling; many of us were sincerely baffled when husbands and boyfriends just didn't get it. Even today, although I have learned that reason doesn't always win the day, I am unable to let go of my faith. I write memos after department meetings, working into the night. I make early morning phone calls. "I'm not sure I fully explained what I meant yesterday. Do you have a minute?" I obsessively revise my academic papers, endlessly honing, elaborating, pruning, clarifying; I feed on the hope that if I can articulate my ideas with precision—anticipating objections, neutralizing potential misunderstanding—I can make feminism convincing.

How did I—growing up with my patriarchal father—come to believe this? Isaac Bashevis Singer's short story "Yentl the Yeshiva Boy" describes a family situation that is not uncommon in Jewish families, especially in families that are without sons:

> [Yentl's father,] Reb Todros . . . had studied Torah with his daughter as if she were a son. He told Yentl to lock the doors and drape the windows, then together they pored over the Pentateuch, the Mishnah, the Gemara, and the Commentaries. She had proved so apt a pupil that her father used to say:
> "Yentl—you have the soul of a man."
> "So why was I born a woman?"
> "Even Heaven makes mistakes."

Letty Cottin Pogrebin, in her angry and honest account of her feelings about her own father, describes a modern-day version of the same:

> He put me through my paces as if he were a rabbi, which he could have been, and as if I were a boy, which I should have been to please him, although he never said it. I adored him for treating me like a son and taking me seriously. He drilled and polished my Hebrew recitation until I was the kind of virtuoso performer that synagogue legends were made of back in 1952, when girls, as a rule, did not do that sort of thing. But I did whatever my father valued. More than anything, I wanted his approval because he was my mentor and I saw myself as his intellectual heir. There was no son to make that claim. . . . Clearly, his legacy was mine if I proved myself worthy of it. Even when I was very young, he made me feel important just by talking to me. He spoke didactically but never condescendingly the way he sometimes addressed my mother and aunts. He talked to me as if I could be trusted to get it on first learning.

Within traditional Judaism, it is sons, of course, who are the preferred recipients of the father's wisdom and knowledge, the sons who
are taught the arts of reason and argument. But even within traditional
Judaism, a father's recognition of a daughter's ability and his own desire to have a challenging conversational partner (or a good listener)
can overcome convention. My own father, unlike Letty Cottin Pogrebin's, did not instruct us in Torah or train us to be intellectual heirs in
any formal way. By the time I was an adolescent, he was usually too
exhausted and depressed to engage in extended conversation with us.
But there had been a time when my father did not return from work
sullen and withdrawn, eager only for his schnapps and an evening in
front of the television, a time when he still saw himself as a scholar
and teacher of his children, a source of wisdom, a communicator of
knowledge, values, ideas.

These memories were, for a long time, overshadowed by my anger
at the way my father had bullied my mother and ignored her needs.
Then one day about eight years ago I noticed how my father, at that
point nearly eighty years old, long since remarried after my mother's
death and the inheritor of several grandchildren by his new wife,
talked to his young granddaughter. He conversed. Really. No baby
talk, no condescending monosyllables of faked comprehension, no distracted nods. His gaze was steady and focused, he listened attentively,
he questioned when he didn't understand. He spoke in full, grammatical sentences. He created the impression that this conversation—about
the rights and wrongs of a minor disagreement that the child had just
had with her brother—was as complex and thought-provoking as a
discussion about the pros and cons of nuclear disarmament. I watched
this conversation, a conversation between equals carried on by an el-
derly man and a five-year-old girl, and I suddenly knew—did not exactly remember, but *knew*—that he had once had such conversations
with me.

I, like Letty Cottin Pogrebin, grew up believing that my father was
passing an intellectual legacy on to me. In my father's case, however,
that legacy was a lack to be filled, ambitions to be carried through to
their rightful conclusion. Later in his life, when his daughters were all
grown up and professionally successful, my father might feign suffering over our unconventional lifestyles and failure to conform to tradi-

tional feminine ideals. But actually he was extremely proud of his brainy girls. "My daughters," he'd moan, shaking his head mournfully but barely suppressing his laughter. "They can't dance. They can't sing. All they can do is think." But smart thinking—and good talking—was what my father admired most of all. A spellbinding storyteller, an avid reader, and an acute analyst of political affairs, he had a wit that was razor-sharp. He like to use it in sneak attack on our expectations of him, and against threatening rifts; he could turn his depressive, self-indulgent moods (and ours) around with the snap of a droll, smart comment. A master Scrabble player, he could do the Sunday *New York Times* crossword puzzle in one morning—in ink.

Contemporary theorists like Daniel Boyarin have admirably tried to construct positive, countercultural images of Jewish masculinity out of strains of Jewish tradition that have shown resistance to aggressive, phallic manliness. For my part, I didn't see much resistance to the phallus in my father's life or in the lives of the Jewish men that I grew up with in the sixties. Many of them were intellectuals and poets, yes; but, like Mailer, Roth, and other postwar Jewish writers, they used their minds like weapons and flexed their words like muscles. Still, I learned those habits too. The taste for disputation and radical critique, tenacity in argument, "muscular" reason—these are forms of phallic power that are easily transferable to women, and that many Jewish fathers have indeed handed over, unstintingly and with great pride, to their daughters. For all our (justified) anger at our fathers, Letty Cottin Pogrebin and I grew up to be feminist gladiators under their tutelage.

There is a price to be paid, of course, for this phallic legacy. In both the Singer story and Pogrebin's reminiscence, the girl blessed with her father's respect and instruction is chosen *because* she is not like other, "real" women. Like them, she has a female body; but unlike them, she has the soul (or mind) of a man. I don't recall my father ever saying this. But he certainly acted as though my sisters and I were of a radically different order than my mother. Like my father, she was good at Scrabble; but judging from the extreme gallantry with which my father complimented her on this, that skill was one of "Heaven's mistakes," an anomaly, a singular and isolated intellectual talent in a soul made for other things. I valued those things: Warmth and sympathy. Cooking and cleaning. Taking care of us. Accompanying him to the track.

Getting her hair done and being taken out to dinner. Being good-humored when teased. But I grew up believing that I had inherited "his" brain.

In fact, in many ways my mother was a far deeper and more perceptive thinker than my father. Like many women (and other groups whose survival depends upon staying in the good graces of the more powerful), she was adept at "psyching" things out, at discerning motivation, need, mood. She was not educated in clarifying the surfaces of things, but she knew how to look behind and under, she could see in the dark. It was she, not my father, who taught me to dig beneath appearances, and to think with my heart as well as my brain. When I was growing up, to have acknowledged any mental kinship with my mother would have been to align myself with those who are not invited into conversations about ideas, who are spoken to condescendingly, whose opinions are not sought. Today, I recognize that my mother's empathic, psychological intelligence is as much a part of how my reason functions as my father's intellectual acuity, and I am profoundly grateful for that.

My father was not a feminist. He rolled his eyes skyward when I lectured my stepmother's son on the sexual politics of Passover seders which ended (as theirs always did) with the women clearing the table and the men watching baseball in the den. My father worried about my politics too. He lay awake for two nights because he had let it slip to a fundamentalist broker that he had a daughter who was a feminist; he was afraid the broker would dispatch right-to-lifers to throw bombs in my window. My father did not have enlightened ideas about men and women. But those conversations I saw him have with his granddaughter and that I know he had with me when I was a little girl—I know with absolute certainty that they were the foundations for *my* feminist self. How different those conversations were from so many other "conversations" I've had at various times with men! Conversations at parties, on dates, at conferences, with men of correct politics and liberal views, men who supported abortion rights (my father never knew that I had had an abortion), believed Anita Hill (my father's first sympathies in such cases always went to the beleaguered man), and taught *Gender Trouble* in their classes (my father believed the differences between men and women were created by God). I think of how many times those liberal men have seemed to look through me as I spoke, or nodded distractedly, or didn't let me get a word in edgewise.

During the Anita Hill/Clarence Thomas hearings I received a call from my father—he was then about eighty years old—wanting to find out my views. This was in itself an unusual occurrence. I suspected what he really wanted was to tell me *his* views, and challenge me to love him in spite of them. He didn't believe Hill, he told me, and suspected that she was being used as a political pawn. Himself a lifelong raconteur of risqué jokes and anecdotes, often told on the job and in mixed company, the concept of verbal offense—so long as the subject was "only" sex—was a stretch for him. He clearly identified with Clarence Thomas and his struggle as a black man to "get respect" from this culture. And, like many men (and some women), he was genuinely puzzled by the fact that Hill had continued to work for Thomas after he had allegedly harassed her. "Why didn't she just quit? Why did she follow him to the EEOC?" he challenged me.

My first impulse was to want to scream, or make some excuse to get off the phone. Those questions were an incessant mantra in the media, and few people seemed to be able to truly hear the answers that were given. My father often shared the editorial perspectives of his favorite male journalists and talking heads, and none of them "got it" at all. What chance was there of this conversation having a happy resolution, in achieving anything other than making me feel crazy? I decided to try anyway; as my father aged, I was inclined to grab whatever opportunities I could to communicate with him. And I had some answers; I had been harassed myself, knew many other women who also had been harassed, and understood that mixture of intimidation, pragmatics, and training in feminine comportment that keeps one from "making a fuss." I swallowed hard, girded my loins like an intellectual gladiator, and tried to explain it all to my father, beginning with my own harassment and how it had made me feel.

After a few minutes, I suddenly sensed that he was listening with the same attentiveness that he had brought to that conversation with his little granddaughter, as he fixed his blue eyes on her and asked her to slow down and explain some notion, important to her but still obscure to him. He asked me questions. He listened to my answers. He interjected, but with genuine curiosity and respect, silently acknowledging my greater expertise and experience. By the end of the phone conversation my eighty-year-old Jewish father, remarkably, had "gotten it." In the midst of that time, when so many of us had begun to doubt that there would ever be any progress made past the stereotypes of lying temptresses and women scorned, when we had begun to despair of the possibilities of communication between men and women, of women's experiences being taken seriously, my father handed me a sweet wild card of hope.

acknowledgments

In 1992, as I was putting finishing touches on *Unbearable Weight*, my book about women and their bodies, I received a letter from Laurence Goldstein, editor of *Michigan Quarterly Review*, inviting me to do a review article on the burgeoning literature on masculinity and the male body. It was as if Larry had read my mind—and hurried its timetable a bit. I had known for a long time that I wanted to write about men and their bodies; it seemed the logical, natural, almost inevitable next step. I just wasn't expecting to begin quite so soon. But I couldn't resist the opportunity, and that article was the beginning of what became this book. So, my first round of thanks goes to Larry Goldstein, for entrusting a review of the literature about men to a woman, and for allowing that woman to be me.

From our very first conversations about this project, my brilliant agent, Beth Vesel, understood with almost uncanny insight what I was hoping to do. Throughout, she has supported my best instincts, pushed my thinking, helped me to focus, got me back on track when I began to wander, kept faith when I faltered. She also facilitated what for me has been a match made in heaven: my relationship with Farrar, Straus and Giroux and my editors Jonathan Galassi and Natasha Wimmer. As I've watched other author friends struggle with the anti-intellectualism of their presses, I have become ever more grateful for the privilege of working with a press that doesn't view "theory" and "general readership" as mutually exclusive terms. Our lives are always informed—although often invisibly so—by ideas. At the same time, ideas are never encountered by living human beings except as they are embodied in experience. I wanted to write a book that would be true to both these facts; if I have succeeded at all, it's been because of the support and guidance of Jonathan and Natasha.

Over the years that I have been working on this book, many colleagues, relatives, friends, and students have generously contributed their own insights, helped with resources, read drafts of chapters, and provided encouragement. Some have made such significant contributions to my thinking—and, sometimes, basic well-being—that "contribution" seems too paltry a word; profound gratitude is more appropriate. Virginia Blum, Leslie Heywood, Binnie Klein, Edward Lee, Ellen Rosenman, and Paul Taylor will all recognize, I hope, where and how their help has shaped this book.

For comments, criticisms, suggestions, stimulating talk, research help at various stages of this project, my thanks to Jonathan Allison, Lynne Arnault, Whitney Baker, Bernadette Barton, Janet Bogdan, Janice Boddy, Barron Boyd, Ron Bruzina, Joan Calahan, Pat Cooper, Stanley Clarke, Tom Digby, Bill Drummond, Nikky Finney, Don Ihde, Don Hanlon Johnson, Patrick Keane, Michael Kimmel, Brandon Look, David Miller, Richard Mohr, Heidi Nast, Dana Nelson, Jill Norton, Susan Silber Nusbaum, Lucius Outlaw, Christian Perring, Suzanne Pucci, Beth Rosdatter, Jennifer Ruark, Jean Rosenblatt, Naomi Schneider, Ron Scapp, Richard Schein, Scott Shapleigh, Jonathan Schonsheck, Marilyn Silverman, Clarence Taylor, Deborah Tooker, Richard Wrangham, and Brenda Weber. My wonderful students in "Philosophy, Masculinity, and the Body," "Masculinity/Femininity: An Interdisciplinary, Multicultural Perspective," "Desire, Knowledge, and the Gaze," and "Gender in Film" have continually provided inspiration, insight, challenge, and great conversation about the topics of this book. My visits to Ohio State University, Grinnell College, Vanderbilt University, Pennsylvania State University, Haverford College, Allegheny College, Miami University, University of Louisville, University of Cincinnati, University of Notre Dame, Yale University, Hamilton College, as well as talks at the Society of Phenomenology and Existential Philosophy, provided invaluable opportunities for discussion as my ideas developed. Unfortunately, it is impossible adequately to acknowledge these sorts of contributions: I can only convey my sincere thanks to everyone who participated.

My gratitude to the wonderful scholars whose written work I have drawn on in this book.

To Linda Wheeler, once again, I couldn't have done it without you.

I am constantly blessed by the humor, intelligence, and love of my family—Binnie, Mickey, Scott, Steve, Estelle, and—always and forever—Edward.

bibliography

in hiding and on display

Beauvoir, Simone de. (1952). *The Second Sex.* New York: Vintage Books.
Bordo, Susan. (1993). "Reading the Male Body," *Michigan Quarterly Review,* Vol. 32, No. 4, Fall (Ann Arbor: University of Michigan), pp. 696–737.
Faludi, Susan. (1994). "The Naked Citadel," *The New Yorker,* September 5, pp. 62–81.
Hollander, Anne. (1994). *Sex and Suits: The Evolution of Modern Dress.* New York: Kodansha International.
Horrocks, Roger. (1994). *Masculinity in Crisis.* New York: St. Martin's Press.
Kessler, Suzanne J., and Wendy McKenna. (1978). *Gender: An Ethnomethodological Approach.* Chicago: University of Chicago Press.
Lord, M. G. (1994). *Forever Barbie: The Unauthorized Biography of a Real Doll.* New York: Avon Books.
Roth, Philip. (1974). *My Life as a Man.* Bantam Books: New York.
Roth, Philip. (1969). *Portnoy's Complaint.* New York: Bantam Books.
Sheets-Johnstone, Maxine. (1994). *The Roots of Power: Animate Form and Gendered Bodies.* Chicago: Open Court.
Solomon-Godeau, Abigail. (1995). "Male Trouble," in Maurice Berger, Brian Wallis, and Simon Watson (eds.). *Constructing Masculinity.* New York: Routledge, pp. 69–76.
Updike, John. (1993). "The Disposable Rocket," *Michigan Quarterly Review,* Vol. 32, No. 4, Fall, pp. 517–520.

hard and soft

Ackerman, Diane. (1994). *A Natural History of Love*. New York: Random House.

Berger, Maurice (1996). "The Mouse That Never Roars: Jewish Masculinity on American Television." In Norman L. Kleeblatt (ed). *Too Jewish? Challenging Traditional Identities*. New Brunswick, NJ: Rutgers University Press, pp. 93–107.

Bernheimer, Charles. (1992). "Penile Reference in Phallic Theory," *Differences*, Vol. 4, Spring, pp. 116–132.

Broder, David. (1998). "Side Effect of Viagra May Be End of a Great Stupidity," *Lexington Herald-Leader*, July 27.

Califia, Pat. (1983). "Gay Men, Lesbians and Sex: Doing It Together," *The Advocate*, Issue 371, July 7, pp. 24–27.

Cameron, Deborah. (1992). "Naming of Parts: Gender, Culture, and Terms for the Penis Among American College Students," *American Speech*, Vol. 67, No. 4, pp. 367–382.

Cowley, Geoffrey. (1998). "Is Sex a Necessity?" *Newsweek*, May 11, pp. 62–63.

Farrell, Warren. (1986). *Why Men Are the Way They Are*. New York: Berkley.

Freud, Sigmund. "Leonardo da Vinci and a Memory of His Childhood." In Peter Gay (ed.) (1989). *The Freud Reader*. New York: W. W. Norton.

Freud, Sigmund. "Lecture XXXII: Anxiety and Instinctual Life." In Peter Gay (ed.) (1989). *The Freud Reader*. New York: W. W. Norton.

Fung, Richard. (1991). "Looking for My Penis: The Eroticized Asian in Gay Video Porn." In Bad Object-Choices Staff (eds.), *How Do I Look?* Seattle: Bay Press, pp. 145–168.

Gilman, Sander L. (1996). "The Jew's Body: Thoughts on Jewish Physical Difference." In Norman L. Kleeblatt (ed.). *Too Jewish? Challenging Traditional Identities*. New Brunswick, NJ: Rutgers University Press, pp. 60–73.

Handy, Bruce. (1998). "The Viagra Craze," *Time*, May 4, pp. 50–57.

Hendren, John. (1998). "Pfizer Presses Insurers on Viagra," *Washington Post*, July 7, p. E3.

Hersey, George L. (1996). *The Evolution of Allure: Sexual Selections from the Medici Venus to the Incredible Hulk*. Cambridge: MIT Press.

Hitchens, Christopher. (1998). "Viagra Falls," *The Nation*, May 25, p. 8.

Irigaray, Luce. (1977). *This Sex Which Is Not One*. Ithaca: Cornell University Press.

Johnson, Susan. (1995). *Brazen*. New York: Bantam Books.

Keen, Sam. (1991). *A Fire in the Belly: On Being a Man*. New York: Bantam Books.

Klein, Joe. (1998). "Primary Cad," *The New Yorker*, September 7, pp. 46–55.

Kulick, Don. (1998). *Travesti: Sex, Gender and Culture Among Brazilian Transgendered Prostitutes*. Chicago: University of Chicago Press.

Laquer, Thomas. (1990). *Making Sex: Body and Gender from the Greeks to Freud*. Cambridge: Harvard University Press.

Lawrence, D. H. (1968). *Lady Chatterley's Lover*. New York: Bantam Books.

Leland, John. (1997). "A Pill for Impotence?" *Newsweek*, November 17, pp. 62–68.

Long, Ron. (1997). "The Fitness of the Gym," *Harvard Gay and Lesbian Review*, Vol. IV, No. 3, Summer, pp. 20–22.

Lopate, Philip. (1993). "Portrait of My Body," *Michigan Quarterly Review*, Vol. 32, No. 4, Fall, pp. 656–665.

Martin, Douglas. (1998). "Thanks a Bunch, Viagra," *New York Times*, May 3.

Morris, Jan. (1974). *Conundrum*. New York: Harcourt Brace Jovanovich.

Mura, David. (1996). "How America Unsexes the Asian Male," *New York Times*, August 22, p. C9.

Pollack, William. (1998). *Real Boys: Rescuing Our Sons from the Myths of Boyhood*. New York: Random House.

Prell, Riv-Ellen. (1996). "Why Jewish Princesses Don't Sweat: Desire and Consumption in Postwar American Jewish Culture." In Norman L. Kleeblatt (ed.). *Too Jewish? Challenging Traditional Identities*. New Brunswick, NJ: Rutgers University Press, pp. 74–92.

Risher, Michael T. (1998). "Controlling Viagra-Mania," *New York Times*, July 20, p. A19.

Safire, William. (1998). "Is There a Right to Sex?" *New York Times*, July 13, p. A21.

Sartre, Jean-Paul. (1966). *Being and Nothingness*. New York: Washington Square Press.

Sharpe, Rochelle. (1998). "FDA Received Data on Adverse Effects from Using Viagra," *Wall Street Journal*, June 29, p. B5.

Steinhauer, Jennifer. (1998). "Viagra's Other Side Effect: Upsets in Many a Marriage," *New York Times*, June 23, pp. B9, B11.

Taylor, John. (1995). "The Third Sex," *Esquire*, April, pp. 102–114.

Theweleit, Klaus. (1987). *Male Fantasies*, Vol. 1: *Women, Floods, Bodies, History*. Minneapolis: University of Minnesota.

Theweleit, Klaus. (1989). *Male Fantasies*, Vol. 2: *Male Bodies: Psychoanalyzing the White Terror*. Minneapolis: University of Minnesota.

Tiefer, Lenore. (1994). "The Medicalization of Impotence," *Gender and Society*, Vol. 8, No. 3, September, pp. 363–377.

does size matter?

Blum, Deborah. (1997). *Sex on the Brain: The Biological Differences Between Men and Women*. New York: Viking Penguin.

Chowdhury, Arabinda. (1996). "Koro: A State of Sexual Panic or Altered Physiology," *Journal of Sex and Marital Therapy*, Vol. 11, No. 2, pp. 165–171.

Chowdhury, Arabinda. (1992). "Psychopathosexuality in Koro Patients," *Journal of the Indian Academy of Applied Psychology*, Vol. 18, Nos. 1–2, pp. 57–60.

Cook, Kevin. (1995). "Is Bigger Better?" *Vogue*, April, pp. 266–268.

Dervin, Daniel. (1995). "The Bobbit Case and the Quest for a Good-Enough Penis," *Psychoanalytic Review*, Vol. 82, No. 2, April, pp. 249–256.

Diamond, Jared. (1997). *Why Is Sex Fun?: The Evolution of Human Sexuality*. New York: Basic Books.

Findlay, Heather. (1992). "Freud's 'Fetishism' and the Lesbian Dildo Debates," *Feminist Studies*, Vol. 18, No. 3, Fall, pp. 563–579.

Ford, Michael. (1996). *Best Gay Erotica 1996*. Pittsburgh: Cleiss Press.

Keuls, Eva C. (1985). *The Reign of the Phallus: Sexual Politics in Ancient Athens*. Berkeley: University of California Press.

Laquer, Thomas. (1990). *Making Sex: Body and Gender from the Greeks to Freud*. Cambridge: Harvard University Press.

Lee, Peter. (1996). "Survey Report: Concept of Penis Size," *Journal of Sex and Marital Therapy*, Vol. 22, No. 2, Summer, pp. 131–135.

Lehman, Peter. (1991). "Penis Size Jokes and Their Relation to Hollywood's Unconscious." In Andrew Horton, *Comedy/Cinema/Theory*, Berkeley: pp. 43–59.

Mailer, Norman. (1998). "The White Negro: Superficial Reflections on the Hipster." *The Time of Our Time*. New York: Random House.

Morgentaler, Abraham, M.D. (1993). *The Male Body: A Physician's Guide to What Every Man Should Know About His Sexual Health*. New York: Fireside Books.

Playgirl, August 1997.

Pronger, Brian. (1990). *The Arena of Masculinity: Sports, Homosexuality and the Meaning of Sex*. New York: St. Martin's Press.

Rowanchilde, Raven. (1996). "Male Genital Modification: A Sexual Selection Interpretation," *Human Nature*, Vol. 7, No. 2, pp. 189–215.

Taylor, John. (1995). "The Long Hard Days of Dr. Dick," *Esquire*, September, pp. 120–130.

Tiefer, Lenore. (1994). "The Medicalization of Impotence," *Gender and Society*, Vol. 8, No. 3, September, pp. 363–377.

what is a phallus?

Awad, George. (1992). "A Fantasy Penis: Its Development, Multiple Meanings and Resolution," *International Journal of Psychoanalysis*, Vol. 73, No. 4, Winter, pp. 661–674.

Bright, Susie. (1991). "What a Friend We Have in Dildos," *The Advocate*, September 10, pp. 66–67.

Califia, Pat. (1983). "Gender Bending, Playing with Roles and Reversals," *The Advocate*, Issue 376, September 15, pp. 24–27.

Cohn, Carol. (1987). "Sex and Death in the Rational World of Intellectuals." *Signs*, Vol. 12, No. 4, pp. 409–439.

Feinberg, Leslie. (1993). *Stone Butch Blues*. Ithaca: Firebrand Books.

Gould, James L., and Carol Grant Gould. (1989). *Sexual Selection: Mate Choice and Courtship in Nature*. New York: Scientific American Library.

Goux, Jean-Joseph. (1992). "The Phallus: Masculine Identity and the 'Exchange of Women,' " *Differences*, Vol. 4, Spring, pp. 40–75.

Keen, Sam. (1991). *A Fire in the Belly: On Being a Man*. New York: Bantam Books.

Lawrence, D. H. (1968). *Lady Chatterley's Lover*. New York: Bantam Books.

Mailer, Norman. (1998). "The Time of Her Time." In *The Time of Our Time*. New York: Random House.

Mohr, Richard, (1992). *Gay Ideas*. New York: Beacon Press.

Pronger, Brian. (1990). *The Arena of Masculinity: Sports, Homosexuality and the Meaning of Sex*. New York: St. Martin's Press.

Simmons, Todd. (1995). "Big Time," *The Advocate*, August 22, pp. 54–56.

fifties hollywood: the rebel male crashes the wedding

Bingham, Dennis. (1994). *Acting Male*. New Brunswick, NJ: Rutgers University Press.

Breines, Wini. (1994). "The 'Other' Fifties: Beats and Bad Girls." In Joanne Meyerowitz (ed.). *Not June Cleaver*. Philadelphia: Temple University Press, pp. 382–408.

Coffin, Patricia. (1967). "A Message to the American Man: Urge You Return to Head of Your Family Soonest," *Look*, January 10, pp. 14–16.

Cohan, Steven, and Ina Rae Hark. (1993). *Screening the Male: Exploring Masculinities in Hollywood Cinema*. London: Routledge.

Douglas, Susan. (1994). *Where the Girls Are: Growing Up Female with the Mass Media*. New York: Times Books.

Ehrenreich, Barbara. (1984). *The Hearts of Men*. New York: Anchor Press.

Fallaci, Oriana. (1967). "Hugh Hefner: 'I Am in the Center of the World,' " *Look*, January 10, pp. 55–57.

Friedan, Betty. (1962). *The Feminine Mystique*. New York: Dell Publishing.

Goodman, Ellen. (1998). "Young's Death Reminds Us We Can't Always Know Best," *Lexington Herald Leader*, July 27, p. A9.

Gray, John. (1992). *Men Are from Mars, Women Are from Venus*. New York: HarperCollins.

Jerome, Jim. (1991). "Riding Shotgun," *People*, June 24, pp. 90–96.

Kael, Pauline. (1994). *For Keeps*. New York: Dutton.

Kael, Pauline. (1980). "Notes on Evolving Heroes, Morals and Audiences," *When the Lights Go Down*. New York: Rhinehart, pp. 195–204.

Kael, Pauline. (1968). *Kiss, Kiss, Bang, Bang*. New York: Bantam Books.

Keough, Peter. (1995). *Flesh and Blood*. San Francisco: Mercury House.

Manso, Peter. (1994). *The Brando Biography*. New York: Hyperion.

Martin, Peter. (1953). "The Star Who Sneers at Hollywood," *Saturday Evening Post*, June 6, pp. 32–33, 88–93.

McCann, Graham. (1991). *Rebel Males: Clift, Brando and Dean*. New Brunswick, NJ: Rutgers University Press.

Pollitt, Katha. (1998). "Women and Children First," *The Nation*, March 30, p. 9 (1).

Strauss, Theodore. (1950). "The Brilliant Brat," *Life*, July 31, pp. 49–58.

Walker, Alexander. (1968). *Sex in the Movies*. Great Britain: Penguin Books.

gay men's revenge

Atwan, Robert, Donald McQuade, and John Wright. (1979). *Edsels, Luckies and Frigidaires*. New York: Delta.

Cavell, Stanley. (1980). *Pursuits of Happiness*. Cambridge: Harvard University Press.

Goodman, Ellen. (1997). "Hollywood's Worry: Can Openly Gay Actor Play a Straight Lead?" *New Haven Register*, August 14.

Groen, Richard. (1997). "Not a Cinderella Story," June 20, My Best Friend's Wedding. [Internet] Http://www.pathcom.com/~shawca/rupert/archive/transcript.html.

Kael, Pauline. (1980). "The Man from Dream City," *When the LIghts Go Down*. New York: Rhinehart, pp. 3–32.

Kempley, Rita. (1997). "Wedding, No Bliss," *Washington Post*, June 20. [Internet] Http://www.pathcom.com/~shawca/rupert/archive/transcript.html.

O'Neill, Tom. (1997). "Crazy About Rupert Everett," *US Magazine*, August. [Internet] Http://www.pathcom.com/~shawca/rupert/archive/transcript.html.

Roth, Philip. (1974). *My Life as a Man*. New York: Bantam Books.

Russo, Vito. (1981). *The Celluloid Closet: Homosexuality in the Movies.* New York: Harper & Row.

Stein, Ruthe. (1997). "Roberts' Act Helps to Keep 'Wedding' Together," June 20. *San Francisco Chronicle*. [Internet] Http://www.pathcom.com/~shawca/rupert/archive/transcript.html.

beauty (re)discovers the male body

Beauvoir, Simone de. (1952). *The Second Sex*. New York: Vintage Books.

Berger, John. (1972). *Ways of Seeing*. Great Britain: Penguin Books.

Blum, Deborah. (1997). *Sex on the Brain: The Biological Differences Between Men and Women*. New York: Viking Penguin.

Boyd, Herbert, and Robert Allen (eds.). (1995). *Brotherman*, New York: Ballantine.

Clark, Danae. (1995). "Commodity Lesbianism." In Kate Meuhuron and Gary Persecute (eds.). *Free Spirits*. Englewood Cliffs, NJ: Prentice Hall, pp. 82–94.

Clarkson, Wensley. (1997). *John Travolta: Back in Character*. Woodstock: Overlook Press.

Ellenzweig, Allen. (1992). *The Homoerotic Photograph*. New York: Columbia University Press.

Farnham, Alan. (1996). "You're So Vain," *Fortune*, September 9, pp. 66–82.

Foucault, Michel. (1985). *The Use of Pleasure*. New York: Vintage Books.

Friday, Nancy. (1996). *The Power of Beauty*. New York: HarperCollins.

Gaines, Steven, and Sharon Churcher. (1994). *Obsession: The Lives and Times of Calvin Klein,*. New York: Avon Books.

Gilmore, David. (1990). *Manhood in the Making*. New Haven: Yale University Press.

Gladwell, Malcolm. (1997). "Listening to Khakis," *The New Yorker*, July 28, pp. 54–58.

Hollander, Anne. (1994). *Sex and Suits: The Evolution of Modern Dress*. New York: Kodansha International.

Long, Ron. (1997). "The Fitness of the Gym," *Harvard Gay and Lesbian Review*, Vol. IV, No. 3, Summer, pp. 20–22.

Majors, Richard, and Janet Mancini Billson. (1992). *Cool Pose: The Dilemmas of Black Manhood in America*. New York: Lexington Books.

Peiss, Kathy. (1998). *Hope in a Jar: The Making of America's Beauty Culture*. New York: Metropolitan Books.

Pieterse, Jan Nederveen. (1990). *White on Black: Images of Africa and Blacks in Western Popular Culture*. New Haven: Yale University Press.

Plato. (1989). *Symposium*. Trans. Alexander Nehama. Indianapolis: Hackett Publishing.

Richmond, Peter. (1987). "How Do Men Feel About Their Bodies?" *Glamour*, April, pp. 312–313, 369–372.

Rotundo, E. Anthony. (1993). *American Manhood: Transformations in Masculinity from the Revolution to the Modern Era*. New York: Basic Books.

Sartre, Jean-Paul. (1966). *Being and Nothingness*. New York: Washington Square Press.

Shaw, Dan. (1994). "Mirror, Mirror," *New York Times*, May 29, Section 9, pp. 1, 6.

Sheets-Johnstone, Maxine. (1994). *The Roots of Power: Animate Form and Gendered Bodies*. Chicago: Open Court.

Spindler, Amy. (1996). "It's a Face-Lifted Tummy-Tucked Jungle Out There," *New York Times*, June 9,

Taylor, John. (1995). "The Long Hard Days of Dr. Dick," *Esquire*, September, pp. 120–130.

White, Shane, and Graham White. (1998). *Stylin'*. Ithaca: Cornell University Press.

gentleman or beast?: the double bind of masculinity

Aristotle. (1941). "On the Generation of Animals." Trans. Arthur Platt, 729a 25–30, p. 676. In Richard McKeon (ed.). *The Basic Works of Aristotle*. New York: Random House.

Bederman, Gail. (1995). *Manliness and Civilization.* Chicago: University of Chicago Press.

Begley, Sharon. (1995). "Gray Matters," *Newsweek*, March 27, pp. 48–54.

Bettelheim, Bruno. (1977). *The Uses of Enchantment: The Meaning and Importance of Fairy Tales.* New York: Vintage Books.

Blum, Deborah. (1997). *Sex on the Brain: The Biological Differences Between Men and Women.* New York: Viking Penguin.

Dean-Jones, Lesley. (1992). "The Politics of Pleasure: Female Sexual Appetite in the Hippocratic Corpus." In Domna C. Stanton (ed.). *Discourses of Sexuality: From Aristotle to AIDS.* Ann Arbor: University of Michigan Press.

Diamond, Jared. (1997). *Why Is Sex Fun?: The Evolution of Human Sexuality.* New York: Basic Books.

Editorial Staff. (1997). "But It's Only Rape: Rally to Aid Tyson Disturbing," *Syracuse Herald-Journal*, February 19.

Gertsman, Bradley, Christopher Pizzo, and Rich Seldes. (1998). *What Men Want.* New York: HarperCollins.

Goldberg, Jonah. (1998). "Boys Will be Boys," *The New Yorker*, February 9, pp. 30–31.

Gray, John. (1997). *Mars and Venus on a Date.* New York: HarperCollins.

Gray, John. (1992). *Men Are from Mars, Women Are from Venus.* New York: HarperCollins.

Guttmacher, Alan. (1987). *Pregnancy, Birth and Family Planning.* New York: Signet Books.

Kantrowitz, Barbara, and Claudia Kalb. (1998). "Boys Will Be Boys," *Newsweek*, May 11, pp. 55–60.

Kimmel, Michael. (1996). *Manhood in America: A Cultural History.* New York: The Free Press.

Lefkowitz, Bernard. (1997). *Our Guys.* New York: Vintage Books.

Oates, Joyce Carol. (1992). "Rape and the Boxing Ring," *Newsweek*, February 24, pp. 60–61.

Pollack, William. (1998). *Real Boys: Rescuing Our Sons from the Myths of Boyhood.* New York: Random House.

Schiller, Lawrence, and James Willwerth. (1996). *American Tragedy: The Uncensored Story of the Simpson Defense.* New York: Random House.

Schulte, Brigid. (1998). "Handsome: Heavy-Duty Hunk or Friendly, Feminine?" *Lexington Herald-Leader*, August 27, p. C16.

Showalter, Elaine. (1997). "Beauty and the Beast: Disney Meets a Liberated Love Story for the 90's," *Premiere*, October, p. 66.

Singer, A. L. (1991). *Disney's Beauty and the Beast.* New York: Disney Press.

Smith and Doe. (1998). *What Men Don't Want Women to Know: The Secrets, the Lies, the Unspoken Truth.* New York: St. Martin's Press.

Stein, Harry. (1994). "The Post-Sensitive Man Is Coming," *Esquire*, May, pp. 56–59.

Warner, Marina. (1994). *From the Beast to the Blonde: On Fairy Tales and Their Tellers.* New York: Farrar, Straus and Giroux.

Wrangham, Richard, and Dale Peterson. (1996). *Demonic Males: Apes and the Origins of Human Violence.* New York: Houghton Mifflin.

the sexual harasser is a bully, not a sex fiend

Some of the material for this chapter is drawn from Bordo, Susan. (1997). "Can a Woman Harass a Man?" *Twilight Zones.* Berkeley: University of California Press, pp. 139–172.

Angier, Natalie. (1995). "Sexual Harassment: Why Even Bees Do It," *New York Times*, October 10, pp. C1, C5.

Bordo, Susan. (1998). "Sexual Harassment Is About Bullying, Not Sex," *Chronicle of Higher Education*, May 1, B6.

Broder, John M. (1998). "Suddenly, Playing Field Is Leveled," *New York Times*, April 2, pp. A1, A16.

Clines, Frances X. (1998). "Paula Jones Case Is Dismissed; Judge Says Even If Tale Is True, Incident Was Not Harassment," *New York Times*, April 2, pp. A1, A16.

Gibbs, Nancy. (1998). "Truth or Consequences," *Time*, February 2, pp. 21–33.

Hill, Anita. (1998). "A Matter of Definition," *New York Times*, March 19, A25.

MacKinnon, Catharine A. (1979). *Sexual Harassment of Working Women.* New Haven: Yale University Press.

Mayer, Jill, and Jill Abramson. (1994). *Strange Justice.* New York: Houghton Mifflin.

Schultz, Vicki. (1998). "Reconceptualizing Sexual Harassment," *Yale Law Journal*, Vol. 107, pp. 1683–1805.

The Starr Report: The Findings of Independent Counsel Kenneth W. Starr on President Clinton and the Lewinsky Affair. (1998). New York: Public Affairs.

Steinem, Gloria. (1998). "Feminists and the Clinton Question," *New York Times*, March 19, A25.

Thomas, Evan, and Thomas Rosensteil. (1995). "Decline and Fall," *Newsweek*, September 18, pp. 31–36.

Toobin, Jeffrey. (1998). "Casting Stones," *The New Yorker*, November 3, pp. 52–61.

Willis, Ellen. (1994). "Villains and Victims," *Salmagundi*, 101/102, pp. 68–78.

beautiful girls, from both sides now

Cleaver, Eldridge. (1991). "On Becoming." In Henry Louis Gates, Jr. (ed.). *Bearing Witness.* New York: Pantheon Books, pp. 176–188.

Farrell, Warren. (1986). *Why Men Are the Way They Are.* New York: Berkley.

Friday, Nancy. (1996). *The Power of Beauty.* New York: HarperCollins.

Kimmel, Michael S. (1991). *Men Confront Pornography.* New York: Meridian.

Morton, Andrew. (1992). *Diana, Her True Story.* New York: Simon and Schuster.

Shweder, Richard A. (1998). "A Few Good Men? Don't Look in the Movies," *New York Times*, January 25, Section 2, pp. 1, 24–25.

humbert and lolita

Abramowitz, Rachel. (1997). "How Do You Solve a Problem like Lolita?" *Premiere*, September, pp. 95–98.

Bohlen, Celestine. "New 'Lolita' Is Snubbed by U.S. Distributors," *The New York Times Book Review*, September 23, 1997.

Bordo, Susan. (1998). "True Obsessions: Being Unfaithful to 'Lolita,' " *Chronicle of Higher Education*, July 24, pp. B7–B8.

Boyd, Brian. (1991). *Vladimir Nabokov, The American Years.* Princeton: Princeton University Press.

Field, Andrew. (1977). *VN: The Life and Art of Vladimir Nabokov.* New York: Crown.

Ingrassia, Michelle. (1995). "Calvin's World," *Newsweek*, September 11, pp. 60–66.

James, Caryn. (1998). "Revisiting a Dangerous Obsession," *New York Times*, July 31, Weekend, pp. 1, 28.

Kael, Pauline. (1962). Lolita, KPFA broadcast, *Partisan Review*, Fall.

Lane, Anthony. "Lo and Behold: Why Can't America See the New 'Lolita'?" *The New Yorker*, pp. 182–184.

Leonard, John. (1997). "The New Puritanism: Who's Afraid of Lolita? (We Are)," *The Nation*, November 24, pp. 11–15.

Nabokov, Vladimir. (1955). *Lolita.* New York: Berkley.

Nabokov, Vladimir. (1961). *Lolita: A Screenplay.* New York: Vintage Books.

Nabokov, Vladimir. (1947). *Speak, Memory: An Autobiography Revisited.* New York: Capricorn Books.

Roiphe, Katie. (1998). "Upfront: The End of Innocence," *Vogue*, January, pp. 38, 40, 46.

Wood, Michael. (1998). "Revisiting Lolita," *The New York Review of Books*, March 26, pp. 9–13.

coda: my father the feminist

Boyarin, Daniel. (1995). "Homophobia: The Feminized Jewish Man and the Lies of Women in Late Antiquity," *Differences*, Vol. 7, No. 2, pp. 41–81.

The Collected Stories of Isaac Bashevis Singer. (1982). New York: Farrar, Straus and Giroux, p. 149.

Gilmore, David. (1990). *Manhood in the Making.* New Haven: Yale University Press, p. 127.

Kimmel, Eric A. (1995). *Bar Mitzvah: A Jewish Boy's Coming of Age.* New York: Puffin Books.

Pogrebin, Letty Cottin. (1996). "I Don't Like to Write About My Father." In Marlene Adler Marks (ed.). *Nice Jewish Girls: Growing Up in America.* New York: Plume, pp. 261–277.